An imprint of The Energy and Resources Institute

© The Energy and Resources Institute, 2016

First published in 2016 by
The Energy and Resources Institute
TERI Press
Darbari Seth Block, IHC Complex, Lodhi Road, New Delhi 110 003, India
Tel. 2468 2100/4150 4900, Fax: 2468 2144/2468 2145
India +91 ▪ Delhi (0)11
Email: teripress@teri.res.in ▪ Website: http://bookstore.teri.res.in

ISBN 978-81-7993-638-2

All rights reserved. No part of this publication may be reproduced, stored in a retrieval system, or transmitted in any form or by any means, electronic, mechanical, photocopying, recording or otherwise, without the prior permission of the publisher.
All export rights for this book vest exclusively with The Energy and Resources Institute (TERI). Unauthorized export is a violation of terms of sale and is subject to legal action.

Author: Shweta Sinha
Publishing Head: Anupama Jauhry
Editorial and Production Teams: Nandita Bhardwaj, Ekta Sharma, Jessica Mosahari; Aman Sachdeva
Designer: Santosh Gautam
Illustrator: Vijay Nipane
Image Researcher: Shilpa Mohan

Printed and bound in India
This book is printed on recycled paper.

CONTENTS

ENERGY – WHAT IS IT?	6
POWER TO THE PEOPLE	10
USING ENERGY	14
RADIANT ENERGY – LIGHT	16
THE MYSTERY OF LIGHT	18
HOW DOES LIGHT BEHAVE?	20
THE MANY COLOURS OF LIGHT	24
MIRRORS AND THE SCIENCE OF REFLECTION	26
EYES AND VISION	30
THERMAL ENERGY – HEAT	32
TRANSFERRING HEAT	36
REAL-LIFE APPLICATIONS OF HEAT	38
LET'S MAKE A SOLAR COOKER	40
WHAT IS SOUND?	42
SOUND AND WAVES	44
ECHO, ULTRASOUND, AND SONAR	46
SOUND POLLUTION	48
LET'S MAKE A MUSIC BOX	50
ENERGY AND THE ENVIRONMENT	52
LET'S GET ACTIVE	54
GLOSSARY	58

ENERGY – WHAT IS IT?

Energy is what drives our world. The sun gives us light and heat during the day. That is a form of energy. At night the bulbs use electricity to produce light. Electricity is another form of energy. Fuel energy runs our cars. We need energy to work and play. It is impossible to think of something that does not use energy. Scientists define energy as 'the ability to do work'.

Energy – Where does it come from?

When you walk or run, you burn energy. Ever wondered how that energy got inside you? It came from the food you ate. We depend on plants and sometimes on animals for our source of energy. So does that mean plants and animals are storehouses of energy? Not really. Plants get their energy from the sun. The leaves contain a green pigment called chlorophyll. Chlorophyll absorbs the energy from sunlight and makes food for the plant through a process called 'photosynthesis'. Humans and animals who feed on the plants, in turn take in the plant's stored energy. When you walk, run or play, your body warms up. It generates heat energy by burning the energy from the food. Energy gets constantly transformed from one form to another in this manner. "Energy cannot be created, nor does it disappear or get destroyed. Energy only transforms from one form to another". This is the 'Law of conservation of energy'.

How was energy first created?

The sun is our primary source of energy. It gives us heat and light. But if energy cannot be created, where does the sun get its energy from? Scientists believe the sun got all of its energy when the Universe was first created billions of years ago. No new energy has been created since or will ever be created again.

Fact file

The sun...
- gives the world as much energy in a minute as we consume in a year.
- has gravitational energy that keeps all planets in orbit.
- is a clean and green source of energy.
- can kill germs in water through UV rays.
- is dying, but we need not worry just yet. The sun's energy will last 6-7 billion years more.

Energy watch — People use energy for everything – from getting out of bed to sending astronauts into space.

Types of Energy

Energy makes us think of things in motion – an exciting football match or of riding a bike. However, even stationary objects contain energy. All energy can be divided into two types – kinetic energy and potential energy. The energy of a body formed due to its motion is known as kinetic energy. The energy that is stored and has a potential to make a change happen is potential energy. A diver standing on a high diving board has stored potential energy because of his position. He uses that stored energy to dive into the water below.

Potential Energy

Kinetic Energy

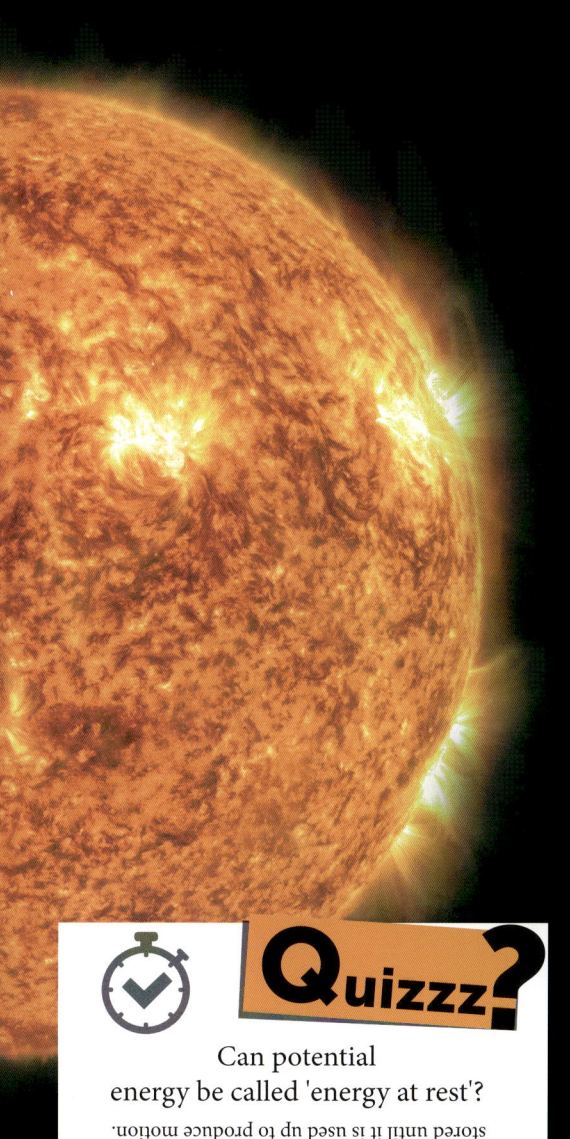

Let's get active

Converting Potential Energy: A catapult
What you need: 7 ice-cream sticks, 3 rubber bands, a bottle cap, cotton balls, glue

1. Stack five sticks together and tie both ends with rubber bands.

2. Take the remaining two sticks and tie one end with a rubber band.

3. Pull the two sticks apart and place the stack of five sticks between these two sticks. Bind the sticks together at the joint with a rubber band. This will hold the catapult together.

4. Stick the bottle cap at the open end of the stick with glue. This is the launching platform.
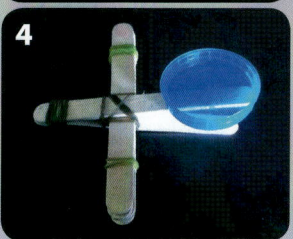

Quizzz?

Can potential energy be called 'energy at rest'?

Answer: Yes, because potential energy remains stored until it is used up to produce motion.

Place a cotton ball on the cap and press the cap down gently. While you hold, the catapult contains potential energy. Release your finger. The catapult springs into action. The potential energy gets converted into kinetic energy, throwing the cotton ball away.

Infographic

One energy, many forms

We experience energy in a variety of ways. Look around you. How many different forms of energy can you spot? Use the chart below as a guide:

Forms of Potential energy

Chemical Energy	 Food	 Fuel	Batteries
Mechanical Energy	 Stretched elastic	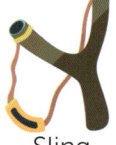 Sling	 A frog about to leap
Nuclear Energy	  Sun and Stars (Fusion)		 earth's core (Fission)
Gravitational Energy	 A roller-coaster cart on top of the hill	 A diver waiting to jump	 Fruit on a tree, waiting to fall

Quizzz?

Which form of energy warms the earth?

Answer: Thermal energy from the sun.

Energy watch — Energy gives us heat, light, and the power to grow and move. Without energy, our world would be cold, dark, quiet and lifeless.

Forms of Kinetic energy

Radiant Energy	 Light	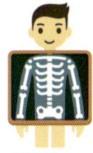 X-Ray	 Laser
Thermal Energy	 Fire	 Heat from the sun	 Steam
Motion Energy	A moving car	A running dog	Ocean waves
Sound Energy	 Talking	 Music	 Animal sounds
Electrical Energy	 Electricity that powers our appliances		 Lightning

Fact file

A roller coaster keeps converting potential energy (gravitational) into kinetic energy (motion) and back as it moves up and down the tracks.

POWER TO THE PEOPLE

What gives an aeroplane the energy to soar the skies? Or ships the energy to cross vast oceans? What makes a kite fly? Electricity powers our appliances, but how do we generate electricity? Will a day come when windmills finally stop moving? We know the sun is a storehouse of energy, but how does the sun sustain its vast energy reserve?

Energy, brought to you by...
Although scientist are always searching for new sources of energy, we depend on nine major ones:

1

Solar Energy
What is it: The sun's energy that has been around for billions of years.
Forms of energy: Radiant, Thermal
Uses: To give light, heat and to generate electricity.

2

Biomass
What is it: Fuel from plants and animals.
Forms of energy: Chemical
Uses: Wood and other fuels burn to give light and heat. Organic waste is used in special waste-to-energy power plants to generate electricity. Animal droppings are converted into biogas. Bio fuel, made from plants like corn and algae, could soon power vehicles.

3

Fossil Fuel
What is it: Fuel from plants and animals that died millions of years ago.
Forms of energy: Chemical
Uses: Coal, crude oil and natural gas power everything from factories to kitchens, and aeroplanes to trains. They are the most widely used energy sources today.

Quizzz?
Candles use paraffin wax as fuel. What kind of energy source is it?
Answer: Paraffin wax is a fossil fuel derived from crude oil.

4

Hydropower
What is it: The power of moving water, like ocean tides, waves, or a river falling off a cliff.
Forms of energy: Motion, Gravitational
Uses: To turn turbines for generating electricity.

Energy watch It is difficult and expensive to store electricity. The supply needs to meet the demand instantaneously.

5. Wind
What is it: Moving air, or wind, containing huge amounts of kinetic energy.
Forms of energy: Motion
Uses: For years, windmills have drawn water from wells, and ground wheat into flour. Wind turbines today generate electricity. One turbine can power up to 300 homes.

6. Geothermal
What is it: The heat from the earth's crust.
Forms of energy: Thermal
Uses: To generate steam from water flowing through underground pipes. The steam turns turbines and generates electricity. Ancient Romans used geothermal energy to heat their homes.

7. Nuclear Power
What is it: The enormous energy stored within the nucleus of an atom. Atoms are extremely small particles, invisible to naked eyes, that make everything else in the universe.
Forms of energy: Nuclear
Uses: Radioactive elements, such as Uranium, are capable of releasing this atomic energy as heat through chemical reactions. Nuclear power plants harness the heat to generate electricity.

8. Hydrogen energy
What is it: Hydrogen gas (H_2)
Forms of energy: Nuclear
Uses: Hydrogen atoms in the sun's core, are constantly combining to form Helium gas. This process, called fusion, releases vast amounts of radiant energy. This is the secret behind the sun's seemingly endless source of energy.

9. Electricity
What is it: Flow of electrical power.
Forms of energy: Electrical
Uses: Electricity lights our homes, keeps it cool, runs televisions and even cars. A world without electricity is unimaginable.

Fact file
In 1904, a farmer in Italy dug a well only to find hot steaming water. The hot water wasn't of much use to him, but soon the idea of generating electricity using the steam was born.

Types of Energy sources

Energy itself cannot be destroyed. However, some sources of energy can get depleted with use, and one day reach an end. Some others are limitless, or can be re-grown and replenished. Energy sources, hence, fall into two categories:

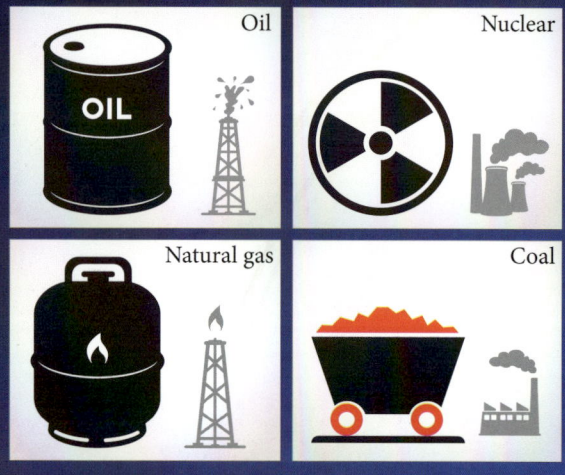

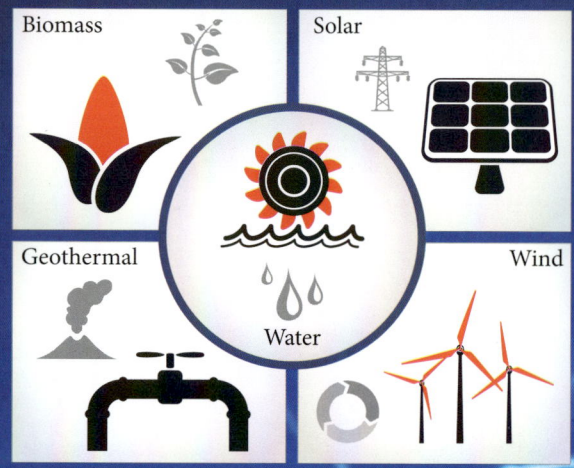

Non-Renewable – Energy sources that have limited quantity.

Renewable – Energy sources that can be recreated.

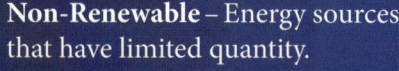 Coal remains the dominant source of energy in our world, followed by oil.

Secondary Energy Sources

Electricity and Hydrogen fall under a third category – 'Secondary Energy Source'. Hydrogen gas (H_2) is lighter than air and easily escapes our atmosphere. This is why Hydrogen exists only in compound form with other elements, such as with oxygen in water (H_2O), and carbon in coal or petroleum. Both electricity and hydrogen are energy carriers. Electricity carries energy produced from other sources such as coal, water, solar etc. In the same way, Hydrogen too is produced from other energy sources.

Quizzz?

Electrical energy can be produced from: mechanical energy or chemical energy?

Answer: Both.

Fact file

India has almost 300 clear, sunny days. The solar power reception is more than enough to fulfil the nation's electricity demand.

Why don't we use more of renewable energy sources?

	The good	The bad
Solar	Free energy	Solar panels are expensive, require wide spaces, and work only while the sun shines.
Biomass	Reuses waste. Plants can be grown again.	Improper implementation leads to deforestation. Burning fuel causes air pollution.
Hydropower	Free energy. Quick response to match electricity demand.	Requires dams to be built, which can be expensive. Dams can result in flooded areas. Located in remote locations, hence cost of transporting the electricity is high.
Wind	Free energy. The land can be used for agriculture.	Unreliable, as it works only as long as there are strong winds. A large number of turbines are required. Located in remote locations. Noisy.
Geothermal	Energy is free and is always available.	Can only be utilized where the hot regions are close to the earth's surface.

USING ENERGY

We use a great deal of energy in our daily lives. At home, in school, in shopping malls – everywhere! Our energy use can be broadly divided into four major sectors – residential, commercial, industrial, and transportation. Out of these, the Industrial sector is the biggest consumer of energy. Moreover, these sectors consume electricity produced by the electric power sector, which itself uses energy.

Residential (homes and apartments)

Electricity has made our lives comfortable, although we use a greater amount of energy now than we did a century ago. Heating and cooling consume the largest share. Next come the electrical appliances – lights, oven, refrigerator, television etc. Water heating is the third biggest consumer. Apart from electricity, natural gas is the most commonly used source of energy.

Modern technology has enabled appliances like washing machines and dishwashers to become energy efficient. Even LED and CFL light bulbs perform better than the traditional incandescent bulbs. Despite these advancements, the energy consumption in homes is increasing. This is because we are plugging in more and more devices than ever before.

Commercial (offices, hospitals, schools, and shopping malls)

Electricity dominates the commercial sector's requirements. Although, needs vary with the type of building, space heating, and cooling consume the most energy. This is closely followed by lighting. Medical equipment in hospitals, computers, and printers in offices, and cooking gadgets in restaurants are other equipment requiring energy. Retail stores and services such as malls, dry cleaners, petrol pumps, together form the biggest consumers of energy. Offices and educational institutes come next.

Industrial (Manufacturing sector)

Every product around you takes up energy to produce. This sector uses energy in a variety of ways. Steel plants require heat to fire blast furnaces and to melt metal. A petroleum refinery requires heat to separate out various products. Heating a boiler to generate steam is yet another example.

Fact file

Until James Watt invented the steam engines in 1780, horses were the primary mode of transport. To describe the strength of his invention, Watt compared the engine's capability to that of a horse's. Thus, the measure 'horsepower' came into being.

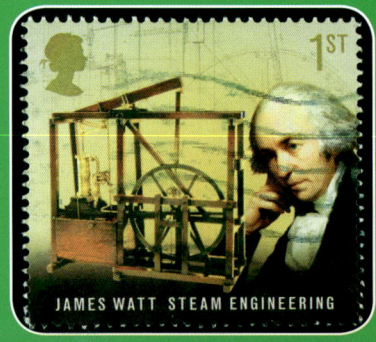

Wood was once the chief source of fuel until coal replaced it in the 19th century. Today, although coal continues to dominate, electricity and natural gas are becoming common too. Manufacturers are also experimenting with other energy sources, like steam, agricultural waste and paper-related refuse. Energy consumption is increasing at a comparatively slower rate in this sector. This is mainly due to adoption of eco-friendly and energy efficient methods.

Transportation

In 2015, it was reported that drivers in the United States of America drove a total of three trillion miles in a single year. Even the earth's circumference is a mere 24,902 miles. And we're talking about just ONE country. Imagine the amount of driving the entire world does. This sector not only consumes a large share of fuel, it also generates a huge amount of Greenhouse Gases. So far, petroleum is the main source of energy. The future will hopefully see vehicles running on eco-friendly bio-fuels.

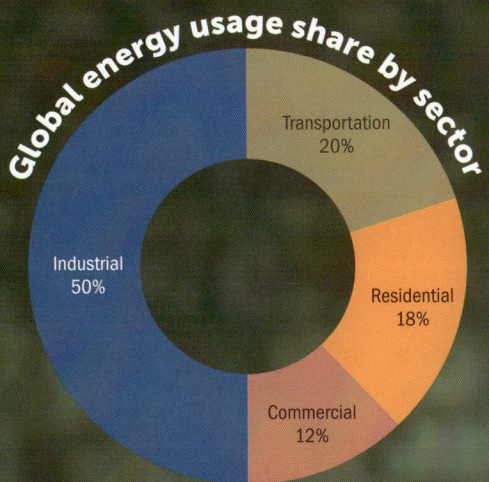

Does our energy use affect the environment?

Answer: Yes, energy sources like coal and petroleum emit harmful pollutants into our environment.

Our Energy usage: Then and Now

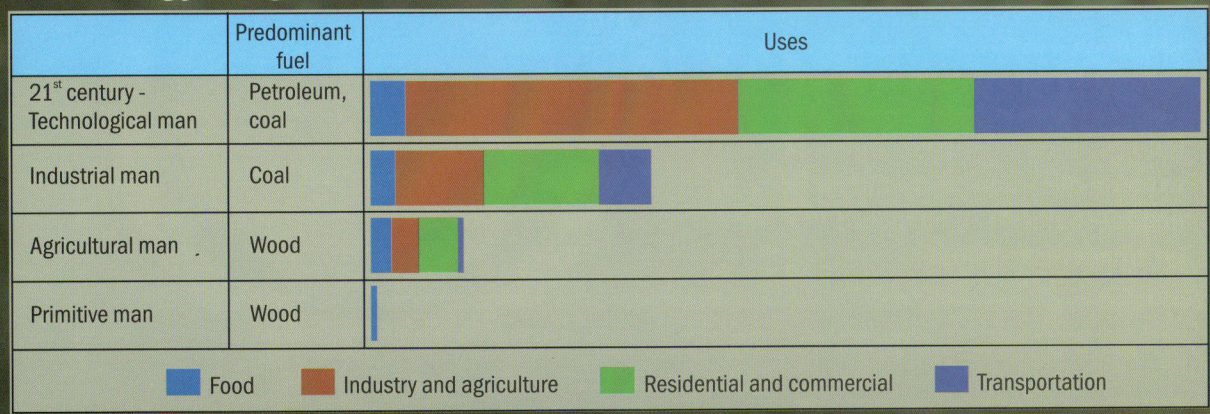

	Predominant fuel	Uses
21st century - Technological man	Petroleum, coal	
Industrial man	Coal	
Agricultural man	Wood	
Primitive man	Wood	

Legend: Food | Industry and agriculture | Residential and commercial | Transportation

Energy watch — The worldwide consumption of energy is expected to grow by more than 40% in the next 20 years.

RADIANT ENERGY – LIGHT

How would it be if the sun refused to rise one morning? Plants would die. The air would turn chilly. The world would become a dead and dismal place. Sunlight not only helps us see, it keeps us alive too. Sunlight keeps our planet warm. It is a source of food for plants. Sunlight is also a source of Vitamin D for us humans.

Sources of Light
Light comes to us either from natural sources like the sun, or from artificial sources, like a candle. Until humans discovered how to make fire, sunlight was their chief source of light.

The Sun
The sun is still our most important source of light. Even a spot as small as a thumbnail is brighter than a million candles burning together. Light from the sun reaches us after a speedy trip of eight minutes. Nothing goes faster than this. However, we receive only a fraction of the sun's light. Most of it flies off into space, away from earth.

The Stars
Like the sun, stars are also made of hydrogen and helium gases. Every second, masses of hydrogen atoms combine to create helium, releasing millions of tonnes of energy. Since the stars are far away, only a minute portion of their light reaches us. Starlight takes several years to make the journey to earth. The star you see today is actually its image from some hundred years ago!

The Moon
The moon does not produce light of its own. Moonlight is simply sunlight bouncing off the moon's surface. It takes 1.255 seconds for moonlight to reach earth.

The light bulb
The light bulb was invented by Thomas Alva Edison in 1879. Since then, they have illuminated our world. The Centennial light in California is the world's longest lasting bulb, burning since 1901. Bulbs work on electricity, and unlike natural sources, can be turned on whenever required.

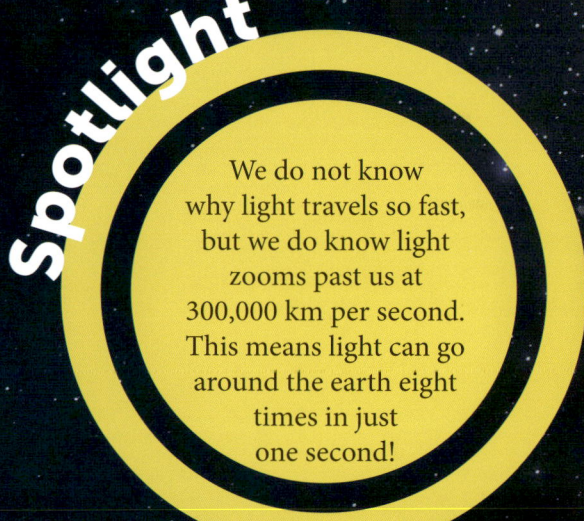

Spotlight

We do not know why light travels so fast, but we do know light zooms past us at 300,000 km per second. This means light can go around the earth eight times in just one second!

Energy watch — LED bulbs are more efficient than incandescent bulbs since they emit mostly visible light and very little heat.

It's hot; No, it's cold

Light is usually produced when something is heated. The sun is a giant ball of fire. The bulb throws light when its filament grows hot. A candle burns to give light. In contrast, cold light or luminescence is produced by a chemical reaction. Fluorescent light bulbs are an example of cold light. They do not generate heat.

Many plants and animals create their own light too. You may have seen blinking fireflies, darting around at night. The Mosquito bay in Puerto Rico is filled with glowing planktons. Most fishes in the deep dark layers of oceans create their own light too. This naturally occurring luminescence is called 'bioluminescence'.

Mosquito bay in Puerto Rico

blinking firefly

What is Light made of?

Light is a type of energy that can be detected by our eyes. But can you touch or feel light? Light may not have physical mass, yet it is made of particles, called photons. The photons travel in tiny wiggly lines or waves that are too small to be seen. If we could make the waves a hundred million times bigger, they would resemble ocean waves. Light is the only stuff that is made of both particles (photon) and waves.

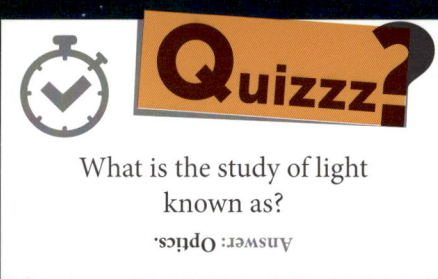

What is the study of light known as?

Answer: Optics.

THE MYSTERY OF LIGHT

Light is pretty impressive. It gets to be the fastest thing on Universe. Nothing can beat the billions of photons rushing from one point to another, creating a beam of light. Light gets to be both a particle and a wave. Although light is mostly white, it can split into different colours and form an awesome rainbow. How do you think light manages to do all of this?

How do atoms create light?

Light is a form of energy and energy cannot be created nor destroyed. So, light too gets created when some other energy is used up. Another form of energy, such as electricity or heat, excites the atoms inside the source of light, such as a bulb's filament. The electrons within the atoms absorb the energy and are kicked farther away from the nucleus. Imagine electrons as being connected to the nucleus with a spring. This extra energy stretches the spring, but makes the atom unstable. The spring needs to recoil into its original state. Hence, the electrons return to their position by giving off the extra energy as photon, or a packet of light.

Inside an atom

Inside an atom lies a centre or nucleus that holds protons, and sometimes neutrons. This nucleus is surrounded by electrons. Atoms look a bit like our Solar System.

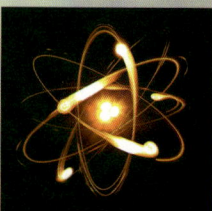

Let's get active

Light walks the straight path

Experiment: The path followed by light is always a straight line, called a ray.

What you need – 3 index cards of the same size, modelling clay, flashlight, hole puncher, ruler

1. Punch a hole in the centre of each card. Draw diagonal lines and punch where the lines meet.
2. Using modelling clay, make stands for the cards. Place the cards upright in a straight line, at equal distances from each other. Ensure the holes line up so you can see through them.
3. Hold the flashlight at the centre of the first card. When you switch on the flashlight, light travels through the three holes and comes out the last one.
4. Now move the middle card so that the holes no longer fall in a line. Notice the beam, it no longer comes out of the third card.

As long as the holes form a straight line, light passes through them and emerges from the end. But when we misalign the holes, light's path gets broken.

Spotlight

Light can make people sneeze when they go from darkness into bright light. This condition, which has perplexed even the scientists, is called a 'photic sneeze reflex'.

The spectrum of light

Light travels in waves. Waves have wavelengths, which is the distance between two neighbouring high points in the wave. Different kinds of lights have different wavelengths. X-ray and UV waves have short wavelengths. They contain high amounts of energy and can harm us. Radio waves and television waves have longer wavelengths. Longer or shorter wavelengths are not visible to our eyes. The range of wavelengths that we see is called 'visible light'. When drawn on a chart, all the wavelengths create an 'electromagnetic spectrum'. The shorter waves lie at one end, the longer ones at the other and the visible light ends up in the middle. Visible light appears white to us, but is a mixture of seven different colours. Each colour corresponds to a different wavelength of light.

Quizzz?

Which electromagnetic wave is known as 'beyond light'?

Answer: Ultraviolet (UV) light.

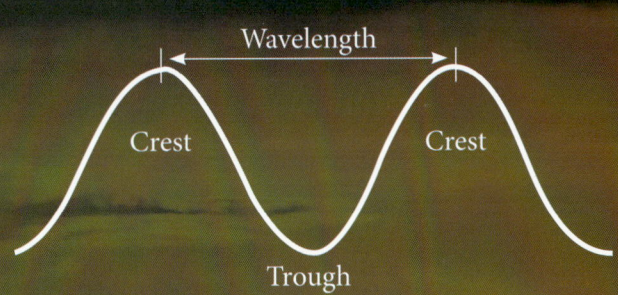

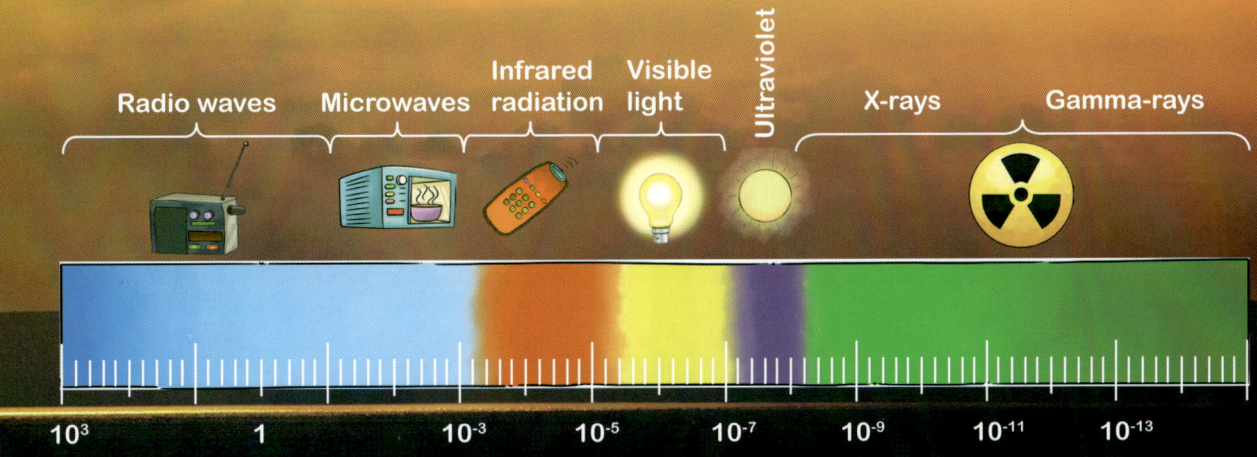

Energy watch

Bees can see ultraviolet light while some snakes can see infrared. Human eyes can see neither.

HOW DOES **LIGHT BEHAVE?**

Light can pass through vacuum. It does not need air or any other medium to propagate. When travelling through empty space, light is at its fastest and always follows the straight path. Interesting things begin to happen when light enters media like water, glass etc. Believe it or not, but light bends, bounces, splits, and basically plays some fascinating tricks.

Light and dark
We see objects only when light bounces or 'reflects' off its surface. If there is no light, we will not see the object anymore. What happens if the object itself comes in the way of light? We see a dark area around the object. This dark area is the shadow and always lies opposite to the light source. A shadow is simply the absence of light.

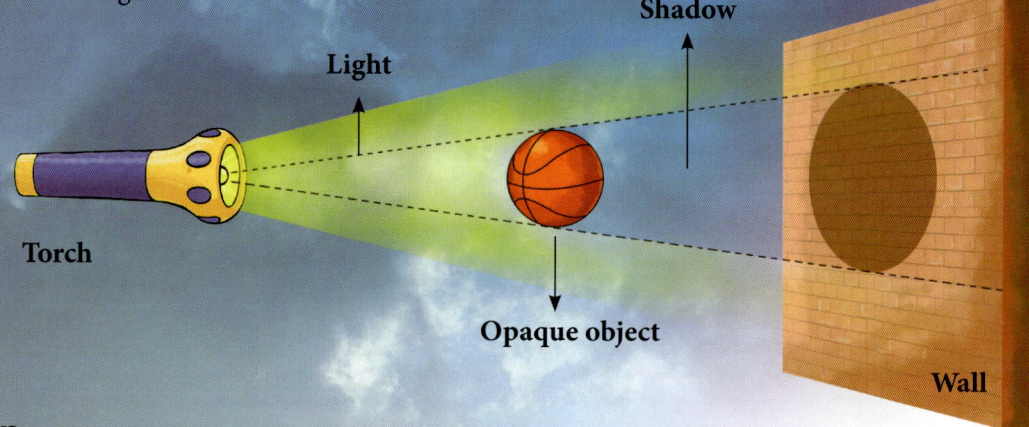

Reflection
Have you ever used a mirror to shine sunlight in another direction? Be careful not to shine it on someone's face. Instead, focus the light on a wall. What you did was change the path of light. Rays falling on the mirror got reflected and began travelling in another direction.
Two types of reflections occur. First is when light falls on a smooth surface, like a mirror or a glossy book cover. This is specular reflection. Specular reflection is uniform and enables us to see our image in a mirror.

The other is diffuse reflection. Diffusion occurs when light bounces off an irregular surface. Most objects aren't highly polished and smooth. In such cases, light scatters all around. Reflection happens, but no clear image is formed.

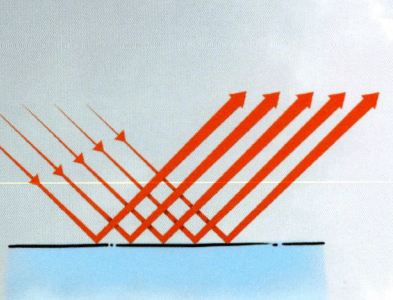

The surface of a road is rough and uneven. Light falling on it gets strewn everywhere. This is diffusion. Observe the same road after a rainfall. Water fills the tiny gaps on its surface. Now light gets uniformly reflected and the road throws a glare. This is specular reflection. Similarly, a clean plate sparkles, but once it is dirty, it appears dull. What kinds of reflection are occurring here?

Refraction

Take a pencil and dip it in water. The pencil appears to bend inside the water. Pull the pencil out and check it. It is dripping wet, but is as straight as it was before. What just happened?

Have you walked in a swimming pool or run into the sea on a beach? Could you walk as fast in water as outside? No, you had to force your legs to move forward. Water is denser than air and even light particles find it tough to push their way through. Same happens with other transparent mediums such as glass and plastics. Hence, after travelling through air, when light enters another medium, it slows down. Due to this, light bends where the surfaces meet. Similarly, light begins to move faster and changes direction when it leaves the medium and enters air. This is why the pencil appears to bend at the surface of water.

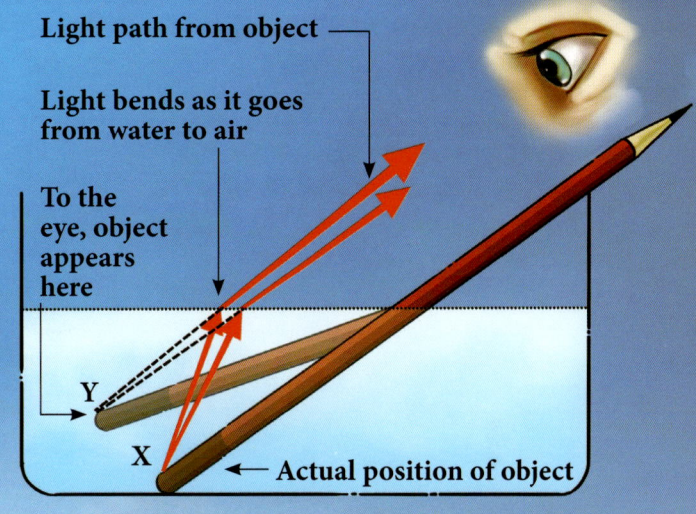

Light path from object

Light bends as it goes from water to air

To the eye, object appears here

Y

X ← Actual position of object

Spotlight

Transparent objects let all light through. Opaque objects block light completely and create clear, dark shadows. Translucent objects block some rays, and let some through.

Quizzz?

Why is it dark in space?

Answer: Because light is visible only when an object reflects it.

Energy watch — Fibre optic cables utilize reflection and refraction of light to transmit data at superfast speeds.

Diffraction

Watch a wave in the sea as it hits a boat. The waves bend around the boat and spread out. All waves, including electromagnetic waves like light, behave in a similar fashion. They bend or get 'diffracted' when they come in contact with an obstacle. Shine a beam of light at the edge of a table in a dark room. Notice that light at the edge spreads out, giving the corner a glow. Diffraction gives clouds their silver lining. Close a door, leaving it open just a slit. Observe the door's shadow and the strip of light streaming through the gap. Is there a sharp line separating the shadow from the light? No, because light bends at the edge of the door and spreads out, blurring the shadow's border.

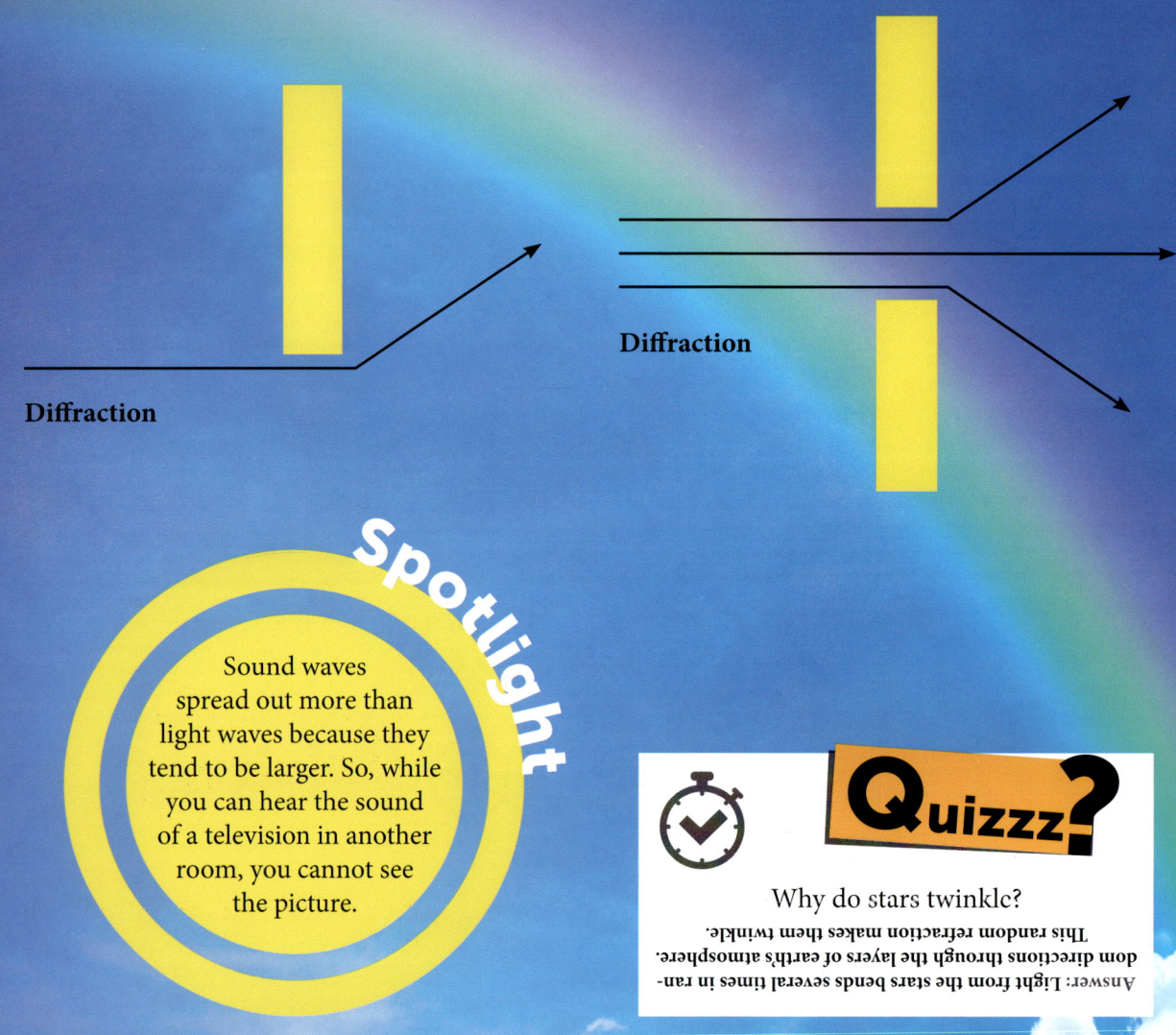

Diffraction

Diffraction

Spotlight

Sound waves spread out more than light waves because they tend to be larger. So, while you can hear the sound of a television in another room, you cannot see the picture.

Quizzz?

Why do stars twinkle?

Answer: Light from the stars bends several times in random directions through the layers of earth's atmosphere. This random refraction makes them twinkle.

Energy watch

Sir Isaac Newton identified the colours of light as VIBGYOR or Violet, Indigo, Blue, Green, Yellow, Orange, and Red.

Interference

Light waves travelling in the same space, sometimes meet and crash into each other. This behaviour is known as interference. When the tops of one wave adds to the other, it is **constructive interference**. Constructive interference increases the intensity of the wave. You can compare this to two people, together pushing a door open. At other times the peak of one wave falls into the trough of the other and they cancel out. We get a flat line instead of a wave. It would be like the two people pushing the door from opposite sides. This is destructive interference.

Why is there a rainbow on a soap bubble?
A soap bubble is a film of soapy-water, surrounding a bubble of air. Light gets reflected from both the outer and the inner surfaces of the soap bubble. After getting reflected, the two rays of light cross each other's path once again. The ray that got reflected off the inner surface travels a greater distance. Hence, the two waves may no longer be in sync. The resulting interference causes some wavelengths to become intense, while others cancel out. Different wavelengths define different colours. So depending on which wavelengths became stronger, the soap bubble gets its colour.

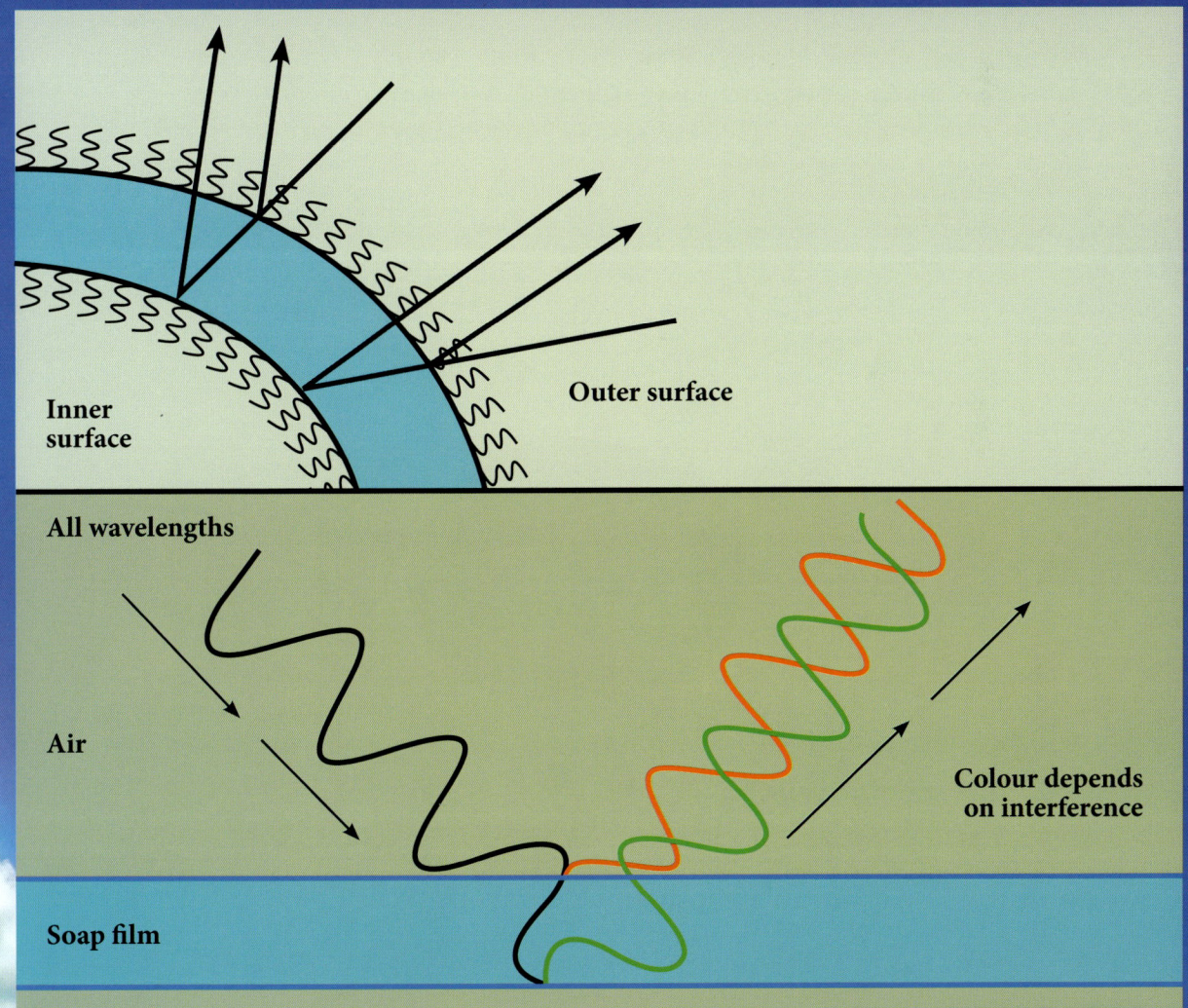

THE MANY COLOURS OF LIGHT

Light itself may not possess any colour, but the world it lights up is tinged in an array of colours. The sky is blue during the day and changes to a bright orange at sunset. Clouds maybe white or grey, yet a rainfall sometimes rewards us with a multi-coloured rainbow. So, how does something appear red, blue, green or even white?

Let's get active

A wonderful, colourful world

Experiment: The colours of light
What you need: Coloured cellophane paper (red, blue, green), flashlight, a dark room
1. Hold the green cellophane paper against the flashlight.
2. Notice, you get green light instead of white.
3. Repeat with other colours and observe the resulting light.

What happened: Visible light consists of seven different wavelengths, each associated with a colour. When visible light strikes an object, certain wavelengths get absorbed, never to be released again. Green cellophane absorbs all wavelengths except green. It lets green to pass through, giving us green light. Similarly, when light falls on a tomato, all wavelengths except red get absorbed. Red bounces off the tomato and reaches our eyes, making it appear red.

Try this: Put a blue filter on the flashlight and shine it on a tomato in the dark room. Does the tomato appear red now?

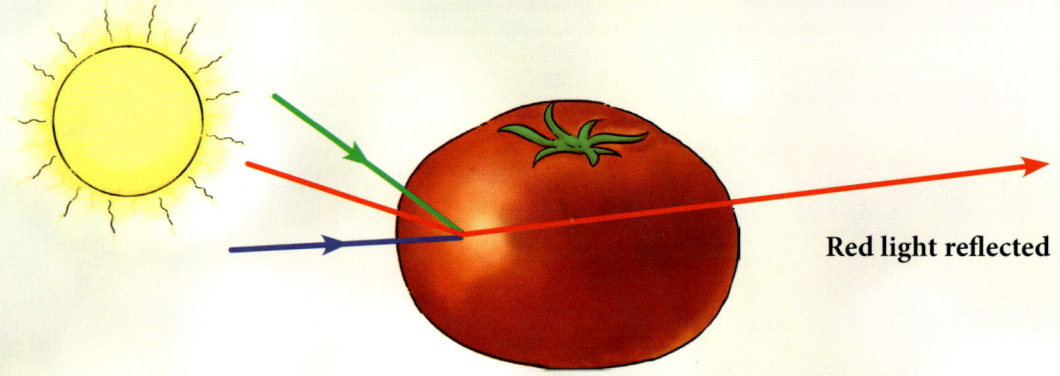

Red light reflected

Blue skies and red sunsets

Gas molecules and tiny dust particles in the atmosphere bounce back blue light more than other colours. Blue light gets scattered and gives the sky its colour. Sunlight is rich in yellow light. As yellow light reaches us directly, without much change, the sun appears yellow during the day. At sunrise and sunset, the sun is low in the sky. Sunlight travels a greater distance to reach us. Most of the yellow and blue gets scattered into the atmosphere by then. What is left is mostly red and orange. This is why during dawn and dusk, the sky around the sun is reddish-orange.

Let's get active

Rainbow science

Why don't rainbows form on dry, summer days? Because a rainbow needs both sunlight and raindrops to combine in a specific manner. When sunlight hits a raindrop at a certain angle, the different wavelengths in it slow down at different rates. As a result, the seven colours refract at varied angles and split, thus creating a rainbow. This splitting of white light into its components is called dispersion.

Experiment: Make your own rainbow

What you need: A shallow bowl of water, small mirror, flashlight, a white sheet of paper, a dark room

1. Put the bowl of water on a table. Place the mirror inside the water at an angle.
2. Ensure the room is really dark.
3. Shine a light on the mirror. Hold the paper in the path of the reflected light and watch a rainbow appear. If the rainbow does not form at first, adjust the angle of the mirror.

Primary and secondary colours of light

White light contains seven different colours. But combining just three – Red+Blue+Green – can give us white light too. These three colours (R,B,G) are known as the primary colours of light. The secondary colours – yellow, cyan and magenta – are produced by combining pairs of primary colours.

Spotlight

Water appears transparent, and not white because it allows all wavelengths to pass through it. It neither absorbs, nor reflects any light.

Quizzz?

Which tool splits white light into its components?

Answer: A Prism.

Energy watch — A white object reflects all seven colours of light, while a black object absorbs it.

MIRRORS AND THE SCIENCE OF REFLECTION

Did you know that glass mirrors were invented only about two thousand years ago? Before that, polished pieces of stone or copper were commonly used as mirrors. The first mirrors perhaps were dark vessels filled with still water. Whatever kind of mirrors people may have used, the science behind them remains the same – that of reflection of light.

All that shines

We see objects either if they generate light (luminous) or if they reflect light (illuminated). The law of reflection states that light is reflected at the same angle it hit the surface. On a smooth surface, parallel rays are all incident at the same angle. Hence, the angles of reflection remain identical, resulting in specular reflection. This law holds just as true for uneven surfaces. Imagine the uneven surface as a combination of numerous tiny smooth surfaces. Each tiny surface differs from others, causing the rays to hit them at varied angles. When angles of incidence differ, their angles of reflection differ too. The reflected rays scatter as a result.

REFLECTION

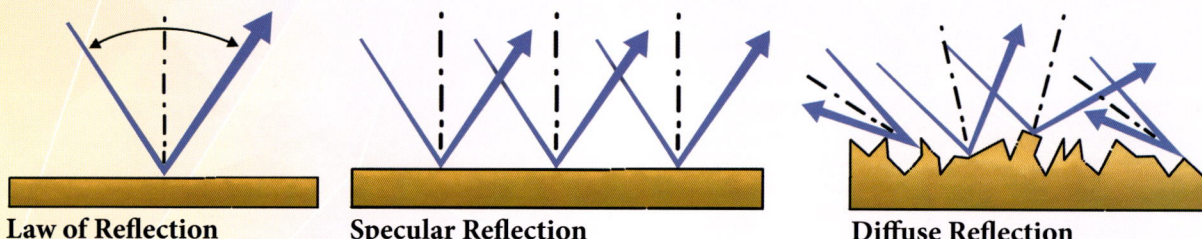

Law of Reflection   Specular Reflection   Diffuse Reflection

Image characteristics

Look at your mirror image and answer these questions:
- Is the image taller or shorter than you?
- Is it inverted?
- Can you get the image on a paper placed in front of the mirror?

Probability is, your image is upright, is your size, and nothing appears on the paper. Images are defined by these characteristics:

- **Real vs. Virtual:** Virtual images are created behind the mirror's surface, at locations where light cannot actually reach. Light only seems to diverge from that location. Real images appear on the same side of the mirror as the object. They can be projected on a screen.
- **Relative size:** Images can get enlarged, reduced or remain the same size.
- **Image reversal:** Some mirrors invert images upside down, while others flip them laterally, i.e. front-to-back.

Energy watch — Not all smooth surfaces act like mirrors as they tend to absorb more photons than they bounce back.

Flipped out!

Look at your image again and raise your left hand. Your image seems to raise its right. Did the mirror inverse your image?

Here's what really happens – Light falls on you and bounces off towards the mirror. The mirror catches this light and throws it back at you. That's how you see your image. A perfect mapping exists between you and your image. If you bring one hand closer to the mirror, the image of that same hand comes closer.

Mirror

You
(looking into the mirror)

Let's get active

What you need: A transparent plastic sheet, mirror, marker.

1. Write 'F' on the plastic sheet.
2. Hold the sheet with the 'F' facing you. The 'F' in the mirror appears correct.
3. Turn the sheet around so the 'F' faces the mirror. The image of 'F' is inverted.
4. Observe the 'F' in the sheet from behind. That looks inverted. Hence, the mirror did not flip the letter – we flipped it from left to right when we turned the sheet around.

But a mirror does flip images laterally –

5. Write 'MIRROR' on the paper and hold it sideways against the mirror. 'M' should be closest to you, while 'R' the farthest. The image shows the opposite. 'R' in the mirror is closest to you while 'M' the farthest.

Quizzz?

Is moon a luminous object?

Answer: Moon is an illuminated object that reflects sunlight.

Spotlight

Mirrors with silver-coated backs, which we use today, were invented in 1835 by a German chemist, called Justus von Liebig.

Curved mirrors

Look into the back of a shiny serving spoon. How does your face appear? Did your nose turn huge? If you've been to a hall of mirrors, you must have seen some oddly shaped mirrors there. The images they throw are funny as hell. You look fat and short in one, or stretched out like an elastic in another. Touch the surface of the spoon or those mirrors. Are they flat? No, they are curved.

A **concave** mirror curves inwards. It is like looking inside a 'cave'. The image characteristics change with the distance from the mirror. Images can be larger or smaller. As objects move farther away, the images turn upside down. They can be both real or virtual depending on the object's distance from the mirror.

If the mirror's surface bulges outwards, it is a convex mirror. Convex mirrors always throw images that are smaller, upright, and virtual.

VIRTUAL AND REAL IMAGES

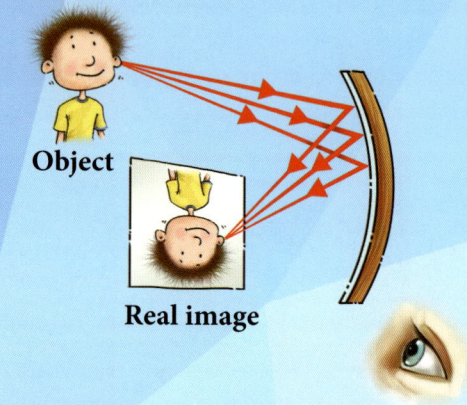

Convex and plain mirrors create virtual images, located behind the mirror.

Real images are created by concave mirrors and can be caught on a screen.

Type of mirror	Application
Concave	Produces a parallel beam of light
Convex	Provides a wider field of view
Concave	Enlarges the image
Concave	Gives off a parallel beam of light that can be dipped or raised
Convex	Allows for a greater view around corners in buildings and streets
Concave	Magnifies the image inside the mouth and helps reflect light

A lens is similar to a curved mirror, only it does not reflect. Instead, a lens lets light to pass through. It is made of transparent material like glass that has curved sides. Magnifying glasses and spectacles are made of lenses.

Eyes in Space

When Dr Lyman Spitzer suggested putting a telescope in space, everyone thought he was joking. Afterall, the year was 1946, and no one had even launched a rocket yet. But, being an astrophysicist, Dr Spitzer felt limited by ground-based telescopes. Even the best of his instruments were dependent upon the earth's atmosphere. The air around us absorbs certain wavelengths like UV and Gamma rays. This results in a huge information loss that the universe sends out. In order to better study space, we needed to be out there.

Dr Spitzer did not give up and finally in 1990, the Hubble Space Telescope (HST) was launched into space. Like other telescopes, the HST also lets in light through an opening at the end of a long tube. The incoming light reflects off a primary concave mirror and onto a secondary mirror. The secondary mirror bounces the light back towards a hole in the centre of the primary. The path of light resembles a 'W'. Behind the primary, several mirrors then reflect and distribute light to a host of scientific instruments.

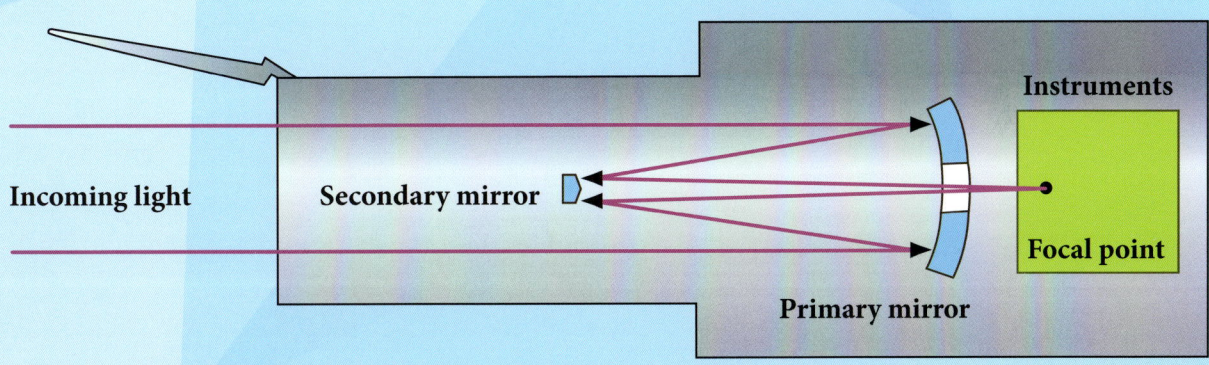

Spotlight

India launched its first space observatory in space on 28th September, 2015. Named the ASTROSTAT, it is a smaller version of the Hubble and weighs 1500 kilograms.

Why is the word "ambulance" written backwards in ambulance cars?

Answer: So that, when seen in a rear-view mirror, the word appears correct.

| **Energy watch** | Only few animals, like the chimpanzees and bottlenose dolphins, are able to recognize their own mirror images. |

EYES AND VISION

Our eyes are an important link to the world. More often than not, we rely on our vision to sense what goes on around us. Eyes are an amazing, one-lens system that help us perceive movements and colours. Animals depend on their eyes not just to look around, but to find food, avoid predators and even to locate a mate.

Looking inside the eye

Your eye is like a ball. Most of the eyeball fits within your skin and muscles. The eye's basic purpose is to collect light.

The eyeball has two parts – a tough white part that protects the eye, and a coloured centre that detects light. The transparent coloured centre is called 'cornea'. Light enters the cornea through a hole (pupil) in the middle. A lens in the cornea focuses the incoming light onto the retina behind. The image that forms here is upside down. Chemicals in the retina convert the image into nerve signals, which are carried to the brain through optic nerves. The brain translates the signals into an upright image, helping us see.

Spotlight

A horsefly's two huge eyes cover most of its head. Each eye contains thousands of lenses that quickly sense movement. That's why a fly moves away faster than you can swat it.

How do eyes detect colour?

The light-sensitive retina contains cones that 'see colour'. Average human eyes have 6 to 7 million cones. That's a huge number, especially considering they are all bunched up on an area as small as a needle point. These cones are not all alike. Some respond to red light, some to green and the rest to blue. Great! But what about the other colours?

Remember, red, blue and green are the three primary colours of light (RGB)? Varying combinations of these primary colours produce the remaining colours. So when you're looking at a banana, wavelengths of yellow light enter your eyes. Inside your eye, both red and green cones respond. Your brain mixes these two responses and tells you the banana is yellow.

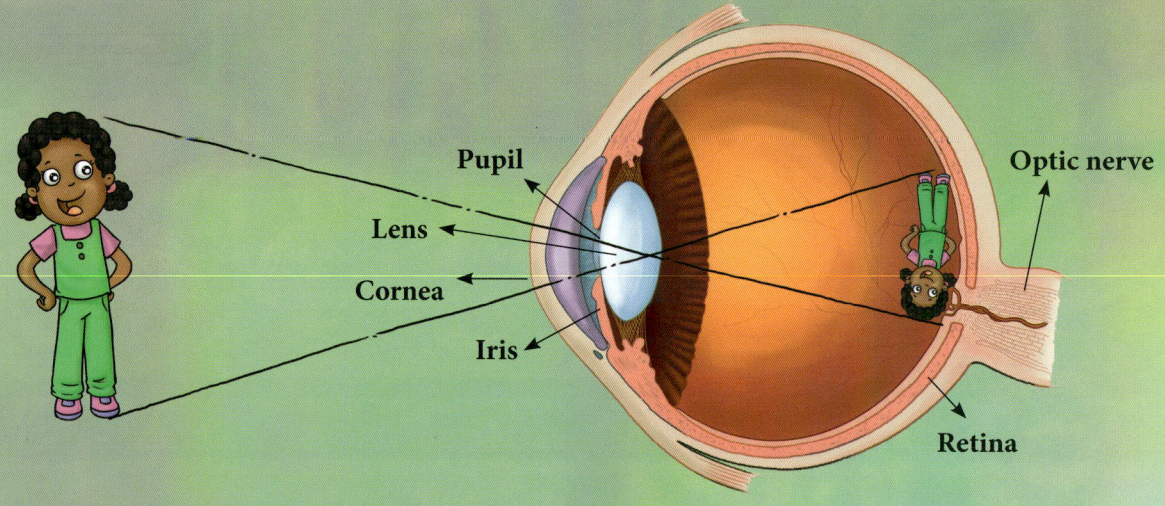

Beware, it's an illusion!

You may have heard stories of thirsty desert travellers getting waylaid by a mirage. They thought they saw water where there was only sand. A mirage is an optical illusion - a trick that light plays on our eyes. When light travels through layers of hot and cold air, it bends. Our eyes seems to see a reflection where there is none This mirage fools us into believing the reflection is a presence of water.

Mirages don't just happen in deserts. When travelling on a hot day, observe the road. The road's surface shimmers as if wet. As you get closer, the 'water' on the road disappears. That's a mirage too.

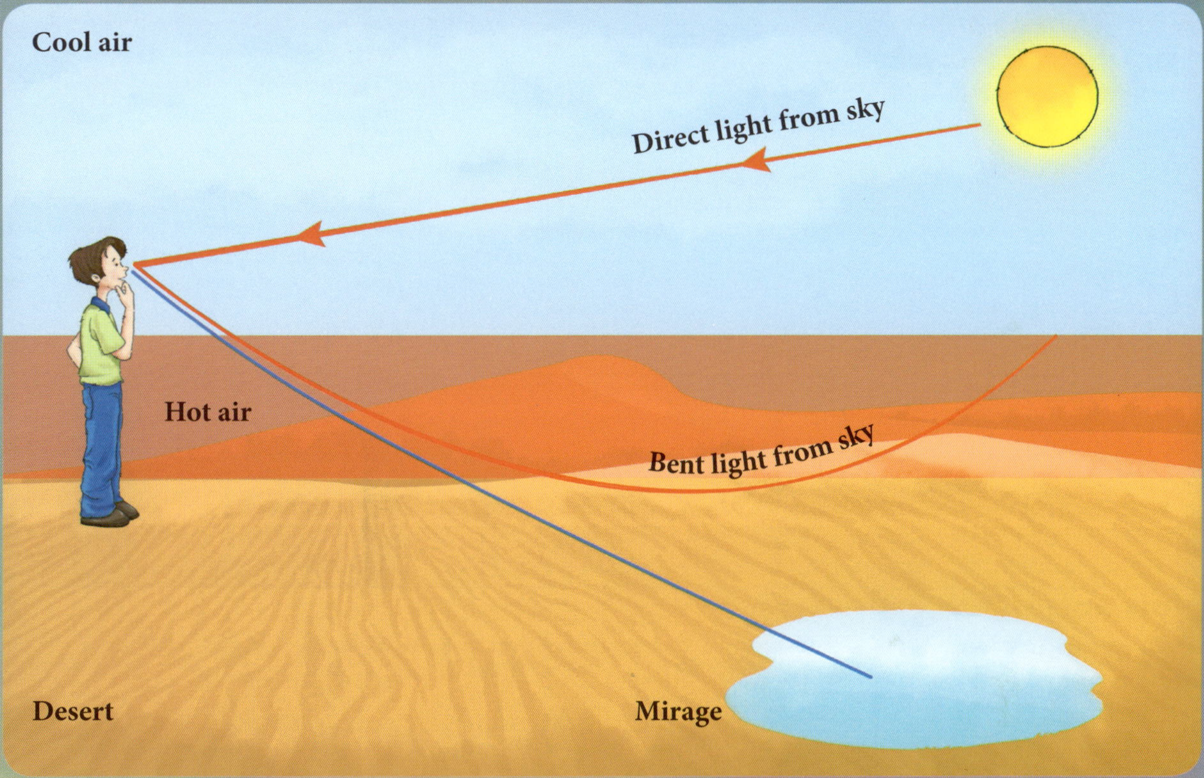

Cool air
Direct light from sky
Hot air
Bent light from sky
Desert   **Mirage**

Seeing in the dark

Tigers, owls, and cats have one thing in common – they all have night vision. This does not mean they can see in complete darkness. Instead, their eyes make the best use of the little light available. Their pupils open wide and round in the dark, letting in as much light as possible. In addition, reflectors behind the retina act like mirrors and bounce light back at the retina. This gives the retina a second chance to sense it.

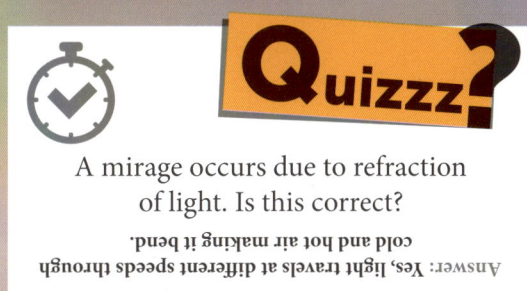

Quizzz?

A mirage occurs due to refraction of light. Is this correct?

Answer: Yes, light travels at different speeds through cold and hot air making it bend.

Energy watch — Two eyeballs help us assess how far an object is. This is called depth perception.

THERMAL ENERGY – HEAT

What does the word 'heat' bring to your mind? Balmy summer days, or sitting around a fire on a winter night? A cup of hot chocolate drink or perhaps a warm shower? How about the kitchen where heat is used in abundance - as fire in the stove, inside the oven or in an induction cook top? Heat is perhaps the most familiar form of energy we see around us.

What is heat anyway?

Heat is the energy that keeps us warm. It is the energy that flows from one object to another when there is a temperature difference. An ice-cube in water will melt. This is because ice, which is colder, absorbs heat from water.

Heat energy results from the movement of molecules within an object. Molecules keep moving all the time and bumping into each other. The hotter the object, the faster they move or vibrate, and higher is their thermal energy. When an ice-cube is dropped in water, energy transfers from water to ice molecules. The thermal energy of the ice-cube increases, causing it to melt. At the same time, water becomes colder as it loses thermal energy.

Joules (J) and Calories (cal) are used to measure thermal energy.

The impatient little molecule

Molecules in solids are tightly pressed together and vibrate at one place. Liquid molecules have more freedom and they like rolling over each other. Gas molecules are the most energetic of the lot, rushing around madly.

How hot?

Temperature tells us how hot something is. We measure it on a scale of degree Celsius. Higher the degrees, the hotter it is. Some people use a different scale called Fahrenheit. To measure temperature, we use a device called a thermometer.

It may seem that objects with a higher temperature contain more thermal energy, but that's not always true. A larger object having low temperature can contain more thermal energy than a smaller object that is hotter. Thermal energy measures the total energy of an object, while temperature is the average energy each particle in the object has.

Temperature vs. thermal energy

	Temperature	Thermal Energy
A pot of boiling water	Higher, as each particle is moving faster than the particles in the iceberg	Lower, as the total number of particles are less.
Iceberg	Lower, as each particle is not moving as fast as the particles in the pot of boiling water.	Higher, as the total number of particles is huge. Even though each particle has low kinetic energy, the sum of all is higher than that of the pot.

Hot and Cold

Touch a cup of tea; it feels hot. Dip your finger in tap water; the water feels cold. But touch an ice-cube for a little while, then dip your finger in tap water. The same tap water feels warm now. Hot and cold are relative terms. Cold is merely the absence of heat. When you touched the ice cube, it sucked in some of the heat and lowered the temperature of your fingertips. In comparison, the temperature of tap water was higher. So, when you dipped your 'cold' finger in the water, it felt warm.

Why don't icebergs melt?

An iceberg is a massive chunk of ice, floating in the oceans near the polar regions. However, icebergs are not frozen seawater. Seawater is salty, whereas an iceberg is made of fresh water. Icebergs were once part of a glacier, but broke off and fell into the water.

Iceberg

Now, water freezes at 0º Celsius. As it solidifies, water molecules line themselves into a crystalline shape. But if there are other particles in water, like salt, they get in the way of the water molecules. As a result, the freezing point of saltwater is lower, letting oceans remain in liquid form even below 0º Celsius. Icebergs floating in these salty waters are surrounded by temperatures below its melting point. So, while the sun does melt the surface, most of the iceberg remains intact for years.

Hot tip

Scientists use a third temperature scale called Kelvin. Absolute zero or 0K is the temperature at which molecules in solids have no energy and stop vibrating completely. This is equal to a freezing -273º Celsius!

Quizzz?

Which will spread faster-drops of colour added to cold water or warm water?

Answer: Colour in warm water spreads faster as the water molecules are moving more rapidly due to heat.

Energy watch

The temperature of a healthy human body is 98.6º Fahrenheit (F) or 37.0º Celsius.

Making heat

The easiest way to make heat is by burning something, like wood or oil. Friction generates heat too. Friction is the force that resists motion between two surfaces. It converts the kinetic energy of motion into thermal energy. Like rubbing your hands warms them up. Even lighting a match depends upon friction. The fuel is in the brown tips of the matchsticks, and friction gives the necessary heat spark to start a fire.

Let's get active

Effects of heat: Getting hotter, getting bigger

Experiment: Make your own thermometer.

What you need: A plastic bottle with lid, scissors, play dough, straw, water, food colouring (optional).

1. Make a hole in the lid using scissors. The hole should be just large enough to let the straw through.
2. Fill the bottle half way with cold water. Add few drops of colour and mix.
3. Screw the lid back. Insert the straw through the hole. Ensure the straw is dipped in the water, but does not touch the bottom.
4. Use the play dough to seal the hole. This will secure the straw and make the bottle airtight.
5. Wrap your palm around the bottle's neck. Watch, as the liquid rises up the straw.

What happened: The heat from your hands warmed the air in the bottle. The warm air expanded and pushed against the water, thus making it rise up the straw.

Like air, even solids and liquids expand upon heating. A traditional bulb thermometer works with liquid instead of gas. Liquid mercury expands and rises up the narrow tube when heated. Similarly, at lower temperatures, mercury cools down and contracts. The level in the thermometer reading thus comes down.

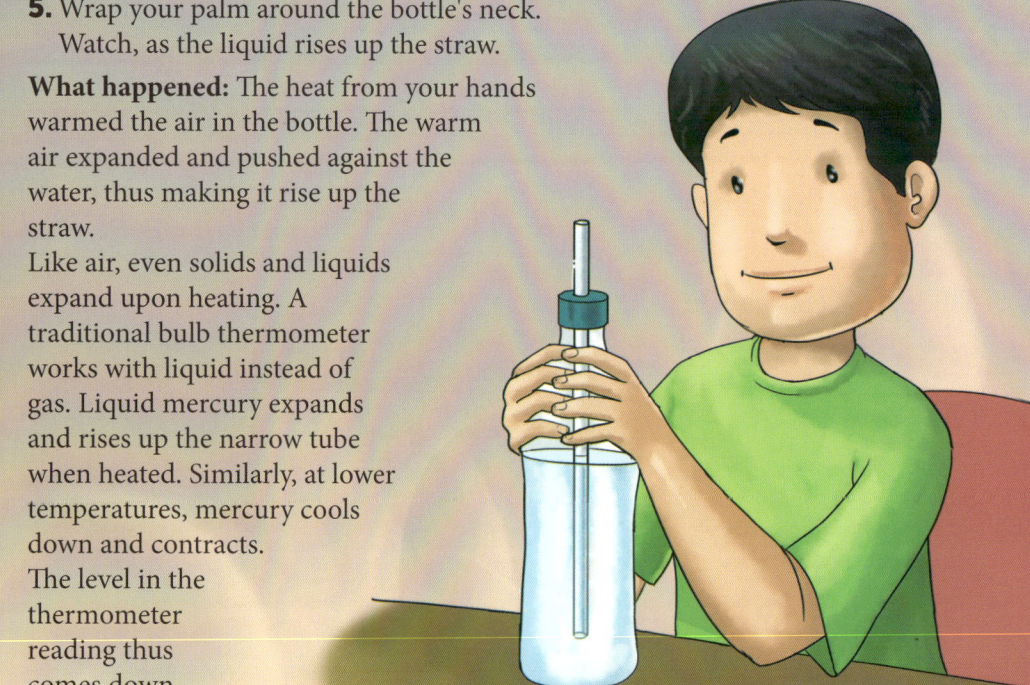

Energy watch

When a gas gives away heat, it 'condenses' into liquid. When a liquid gives away heat, it 'solidifies'.

Let's get active

Effects of heat: Shape shifting
Experiment: Heat turns solids into liquid, and liquid to gas.
What you need: Ice cubes, cup

1. Leave the ice cube in the cup.
2. In a little while all of the ice melts into water. Run a finger above the surface of water; it feels cold. The ice cube absorbed the heat from the air around it and melted into water.
3. Take some warm water and dip your finger in it. Hold your finger up. Preferably, stand next to an open window.
4. In a little while your wet finger becomes dry and your finger feels cold. The water sucked up heat from your finger and evaporated.

What happened: Solids take in heat and melt into their liquid state. The temperature at which a solid melts is known as its melting point. Melting point is different for different solids. Similarly, the temperature at which liquid bubbles and boils when heated is known as its boiling point. Molecules of liquid vibrate faster as they absorb heat energy. Some of the liquid molecules get so excited, they escape the liquid's surface and turn into gas even before the liquid starts to boil. This is called evaporation.

Hot tip

Solid gold melts to liquid at 1064° C and to gaseous form at 2856° C. Compare that to water, which melts at 0° C and boils at 100° C.

Why are there gaps between sections of railway tracks?

Answer: To give the metal tracks space to expand during hot weather.

TRANSFERRING HEAT

We feel warm in the sun because the sun's heat transfers to our bodies. We can cook because heat transfers to the utensils and to the food in it. What if heat got lazy and refused to move? Things would then remain either hot or cold! Thankfully, heat likes to mix around in not just one, but three different ways.

Conduction

Dip a spoon in a cup of hot chocolate. In a little while, the handle of the spoon becomes hot too. The particles in the spoon's surface inside the liquid gain energy and begin to vibrate faster. These particles or molecules are packed tightly inside the solid metal, so they do not move around. Instead they vibrate at their own position. As they wiggle with energy, they knock against neighbouring molecules. The neighbouring molecules begin to vibrate faster too and bump into other molecules. Thermal energy, thus passes from one vibrating molecule to the next until it reaches the other end. This process is called conduction.

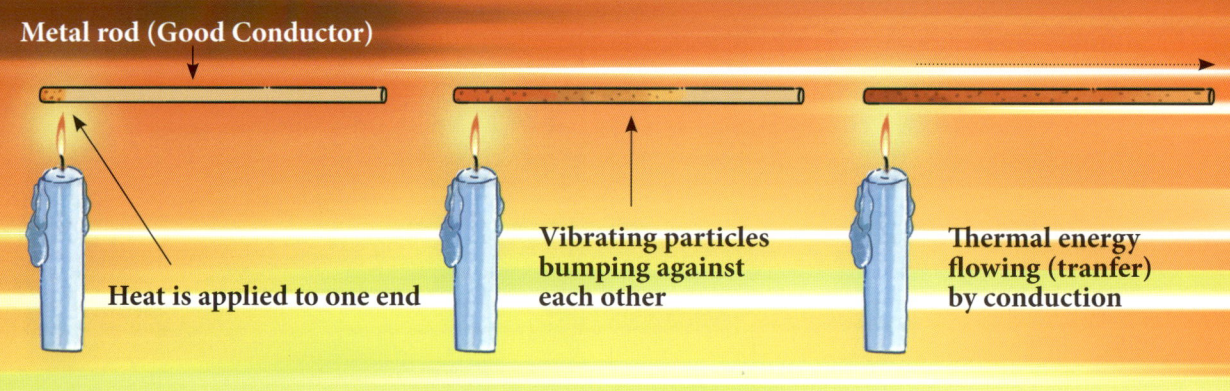

Conductors and Insulators

Conduction works best in solids where molecules are tightly lined up. Metallic solids are good conductors of heat. However, not all solids allow thermal energy to pass through easily. Plastic, wood and cloth trap heat and are poor conductors or 'insulators'.

Can you label the items?

	Conductor or insulator?
i. Copper pipes	
ii. Oven mitt	
iii. Sweater	
iv. Baking sheet	
v. Bottom of a frying pan	
vi. Handle of a frying pan	

(Answer key: Conductors - i, iv, v. Insulators - ii, iii, vi)

Convection

When a large number of heated molecules move from place to place, they carry thermal energy with them. This process of transferring heat from one area to another is called convection. Convection requires free movement of heated particles. Such movement is possible only in liquids and gases, where molecules are loosely bound to each other.

Sea breeze is a result of convection currents. In coastal regions, the land heats up faster than the sea during the day. As a result, hot air above the land rises up and cool air from the sea rushes in to replace it. This breeze from the sea to the land is called sea breeze. At night, the opposite happens and is called the land breeze.

Energy watch — An igloo is a house made of ice, but it keeps the insides warm because ice is a good insulator.

Hot tip
When hot and cold items come in contact, the hotter object transfers its heat to the colder one, and cools down. Ultimately, the heat flow stops when both reach the same temperature.

Why do cooking utensils have plastic or wooden handles?

Answer: As wood is a poor conductor of heat, the handle remains cool even while cooking.

Radiation

Heat can transfer without involving any particles at all. The heat from the sun travels through vacuum in space to reach earth. Heat piggybacks on electromagnetic waves to get transferred. In a space heater, the rod is heated and its heat gets radiated around. This can easily be seen as the rod glows red-hot.

Energy likes to move

When we make a fire by burning logs, all three ways of heat transfer take place. Heat transfers from wood to wood through conduction and keeps the fire going. The air above the fire heats up and rises, creating convection currents. And the heat that we feel directly from the fire, reaches us through radiation.

REAL-LIFE APPLICATIONS OF HEAT

Heat keeps the cold away, but that's not all there is to it. Heat has no mass; it is not a physical substance. However it has properties that can be utilized to do work in many wonderful ways. Until the steam engine was invented, horses pulled carriages, sails drove ships and oxen ploughed fields. The steam engine managed to harness fire to replace older methods of doing work.

Specific heat capacity

Heat is needed to cause any change in temperature. If the particles of a substance are hard to move, then more energy is required to make the change. Particles that move easily require smaller amounts. How much heat is required, depends on :
- Substance being heated
- Quantity being heated
- By how many degrees does the temperature change

When the degree rise is 1° C, the amount of energy used is called the **heat capacity** of the substance. A low heat capacity means little energy is needed to heat the substance. To get a useful comparison between substances, scientists use equal quantities. Hence, the **specific heat capacity** is the amount of energy needed to change the temperature of one unit (1 kg or 1 gram) of the substance by *one degree Celsius (1° C)*. This applies to both cases –
i) When heat is added to raise the temperature
ii) When heat is taken away to lower the temperature

Specific heat capacity: Let's compare

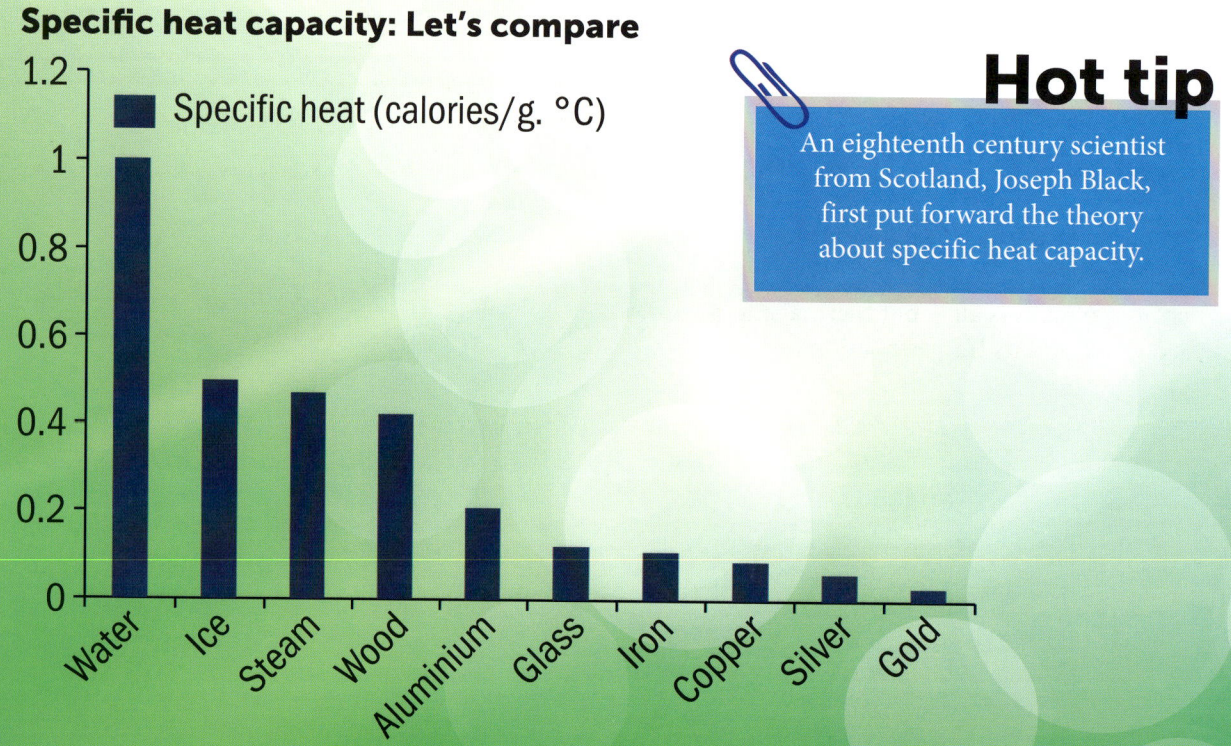

Hot tip

An eighteenth century scientist from Scotland, Joseph Black, first put forward the theory about specific heat capacity.

Letting heat do work

Heat is a bundle of energy. A heat engine takes in this energy and uses it to perform work. They utilize the temperature difference between a hot 'source' and a cold 'sink' to convert the heat energy into mechanical energy.

- **Steam engine** – A steam engine is a vapour-powered engine. Coal is burned to provide the heat input to the engine. The heat boils water to generate steam. This expanding steam is captured and is used to push the engine forward (work output). Any excess heat is rejected through an exhaust.

Try this:
To see the power of steam, boil water in a kettle and cover it loosely with a lid. When the water begins to boil, the rising steam pushes the lid off the kettle.

- **Refrigerator** – A refrigerator is a heat pump, or a heat engine in reverse. Instead of 'work' as an output, work is done to move heat. The external power source (electricity) provides the work-input to the refrigerator. Extra heat is absorbed from the cold space (refrigerated compartment) to keep it cold. This heat is pumped out to the 'condenser' at the back. From here, heat is released into the surrounding air (hot space).

Thermodynamics

Thermodynamics is all about thermal energy: how heat is used, and its transformation into other forms of energy. The sun heats the earth – its land, oceans, and the air. Giant air and water currents swirl around the planet as a result of this heat. This movement of air and swirling of water happens because of transformation of heat into work. This is an example of a thermodynamic system in nature. Can you spot other such systems around you?

Are heat engines a type of thermodynamic system?

Answer: Yes, all heat engines are types of thermodynamic systems. They convert thermal energy into mechanical energy.

Energy watch — Our universe is the largest thermodynamic system. When the universe ends and all its energy dissipates, so will thermodynamics.

LET'S MAKE A **SOLAR COOKER**

A solar box cooker cooks food by trapping the sun's heat in an insulated box. Solar cooking has helped people in several developing nations to reduce their demand for firewood. Also, it provides a clean, smoke-free environment for cooking. While some solar cooker models are available in the market, you can make a small one easily at home.

To make a Solar Box Cooker, you will need
- A Cardboard box with a lid (use a pizza box or a shoe box)
- Scissors / Craft knife
- Roll of aluminium foil
- Clear plastic sheet or cling film
- Black chart paper
- Bamboo stick or a wooden scale
- Newspaper
- Glue, tape

1. Measure a rectangle on the lid, leaving a 1-inch gap from the sides. Cut three side, leaving one longer side uncut.

2. Cover the flap with aluminium foil by tightly wrapping it around the cardboard and taping it to the back. Fold this flap out so it stands when the lid is closed.

3. Use clear plastic sheet to create a window for the sunlight to enter the box. Do this by opening the lid and lining it from inside. Tape the sides securely to the lid, sealing out the air.

4. Cut out black chart paper and line the inside of the box with it. Black colour absorbs heat. The food will be placed on this black surface.

5. If you want, you can add insulation by rolling up sheets of newspaper and lining the box. This will help hold more heat inside the box. Make sure to secure the rolls of newspapers with tape first.

6. Close the lid and keep the flap open using a bamboo stick, wooden ruler or a pencil. Your Solar cooker is ready!

7. You can toast bread by buttering up a slice and placing it in your oven in the sun. Focus the sun's rays on the toast by adjusting the flap.
   Be careful while taking out the toast. It will be hot!

Energy watch Even as early as in 700BC, people made fire using solar energy and a magnifying glass lens.

WHAT IS **SOUND**?

Trrrriiiinngg... your alarm rings. You mother calls out to wake you up. A bird tweets on your window sill. A car honks on the road even as a dog barks back at it. The kettle whistles in the kitchen while someone sings on the radio. Sounds begin to surround you even before you begin your day. But what exactly is sound? Let's find out.

Sound and vibration

Pluck the string of a guitar. Can you see it moving back and forth? It is this back and forth motion that produces sound. All sounds are generated by vibrations. Some vibrations are easily visible, like the guitar's string. Others you may not see, but can feel them. Touch your throat and hum. You will feel gentle vibrations. Tap a table. You can hear a quiet sound, but the vibrations are so mild that you can neither see, nor feel them.

How does sound look like?

Sound is an invisible form of energy, created by fast, tiny movements. When something vibrates, it pushes the air particles around it, making them vibrate too. The particles push neighbouring particles and the vibration moves forward. The particles vibrate in a back and forth motion. As they move forward, they push other particles, crowding them together. As they move backwards, the space gets less crowded. The vibrations ripple forward as a chain of alternating crowded and not-so-crowded areas. This ripple forms the sound wave. Sound waves can travel through air, liquid and solid, but not through empty space.

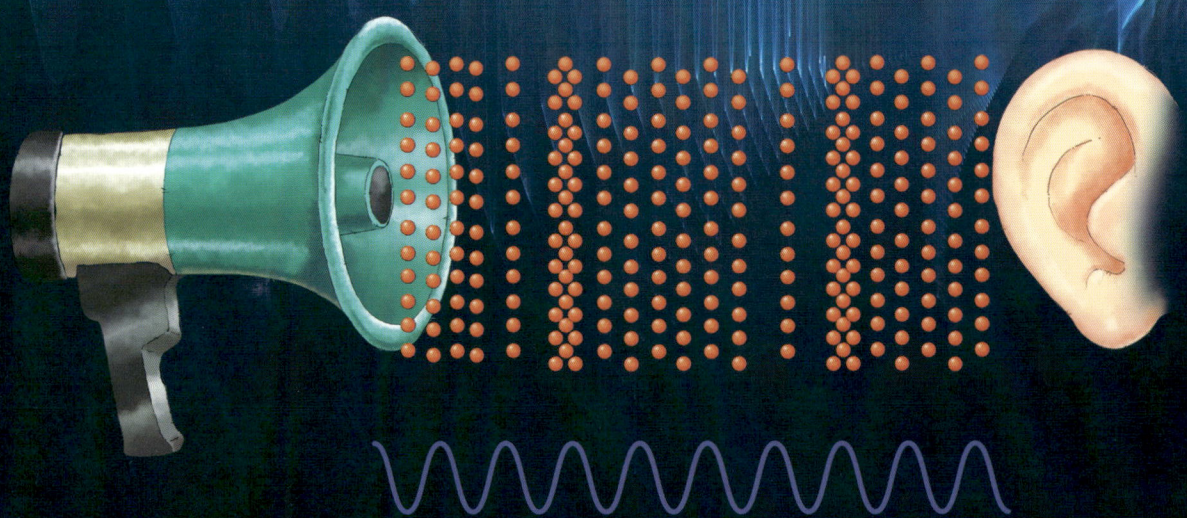

Energy watch — Sound travels four times faster in water than in air, at a speed of 1,482 meters/second.

42

If a spaceship blew up in space, would you hear it?

Answer: No, there will be no sound as outer space is vacuum.

Extremely loud sounds, whether music or noise, create such high pressures that they damage human ears.

Sounds good

Let's get active

Experiment: Sound travels through solid
What you need: Two plastic cups, a long piece of string
1. Make holes in the bottom of both cups.
2. Insert the string through the holes, so the backs of the cups face each other.
3. Knot the loose ends of the string.
4. Have a friend take one cup, while you carry the other. Stand far apart. Ensure the string is stretched and doesn't touch anything.
5. Hold your cup to your ear. Ask your friend to speak softly into his cup. Can you hear his voice?

The string of your 'telephone' carries the sound to your ear.

The Speed of Sound

Sound travels through air at a speed of 343 metres per second. Sound travels faster in water and even faster in steel. The speed of sound depends on the type of medium. The closer the particles within the medium, the faster the vibrations get transmitted.
Something moving at the speed of sound, is said to be travelling at Mach 1. When a jet flies faster than this, it breaks the 'sound barrier'. As it surpasses the barrier, the jet causes a sonic boom that sounds like a clap of thunder. Sometimes, the jet sheds the water droplets that condense on its body, creating a white mist halo of vapour around it.

Amplitude and frequency

Amplitude of a wave defines how big it is - how high is its peak. Stronger waves push particles with a high amount of energy and give taller amplitudes. Frequency is the number of times the wave goes back and forth in one second. Frequency is measured in Hertz (Hz), or cycles per second. It determines whether we can hear the sound or not. The audible range of human ears falls between 20Hz and 20,000 Hz. However, we hear best between 1000 - 5000 Hz, around which a normal conversation is centred.

SOUND AND WAVES

The science of sound is called acoustics. But, why the need to study sound? Understanding sound has many uses. Doctors use sound to see inside the human body. Architects need to ensure theatres allow sound to travel smoothly to everyone. Musicians must have a sound knowledge of their instruments. Without understanding sound, we would have no radios, televisions or even telephones!

The nature of sound

A melodious song is pleasing to hear while the traffic's noise annoys us. What makes each sound different?

Pitch: Compare a lion's growl to the squeak of a mouse. While one is deep and low, the other is shrill and high. This property, called pitch, depends upon the wavelength of the sound wave. Wavelength is the distance between one peak and the next. Shorter the wavelength, higher the pitch.

Loudness: Loudness is the most obvious property of sound. The amplitude of the wave decides how loud the sound is. Bigger the sound wave, the more air pressure it creates and the louder it gets. Loudness is measured in decibels (dB).

Timbre: What makes your voice unique from your friend's? Why does a guitar sound different from a drum playing the same tune? The reason is, most sounds are a mix of multiple waves, which combine to give sound its uniqueness. Timbre is the quality that distinguishes one sound from others.

Loudness of various sounds

Normal conversation:	60 dB
Violin:	84 - 103 dB
Jet engine at 100 ft:	140 dB
Rock music peak:	150 dB
Loudest sound possible:	194 dB

Sounds good

Why do you see lightning before you hear the thunder?

Answer: Because light travels much faster than sound.

How do we hear?
Our ears are so fine tuned, they can process information a thousand-times faster than our eyes. Sound waves are captured by our outer ear and are transported to the ear drum inside, causing it to vibrate. The ear drum passes the vibrations to three connected bones called the hammer, anvil and stirrup. These bones in turn tap the oval window of the inner ear that is filled with fluid. The fluid begins to vibrate with these bones. They help the hair cells or sound receptors to convert the vibrations into electrical signals, which flow into the brain. The brain interprets these signals and we "hear" the sound.

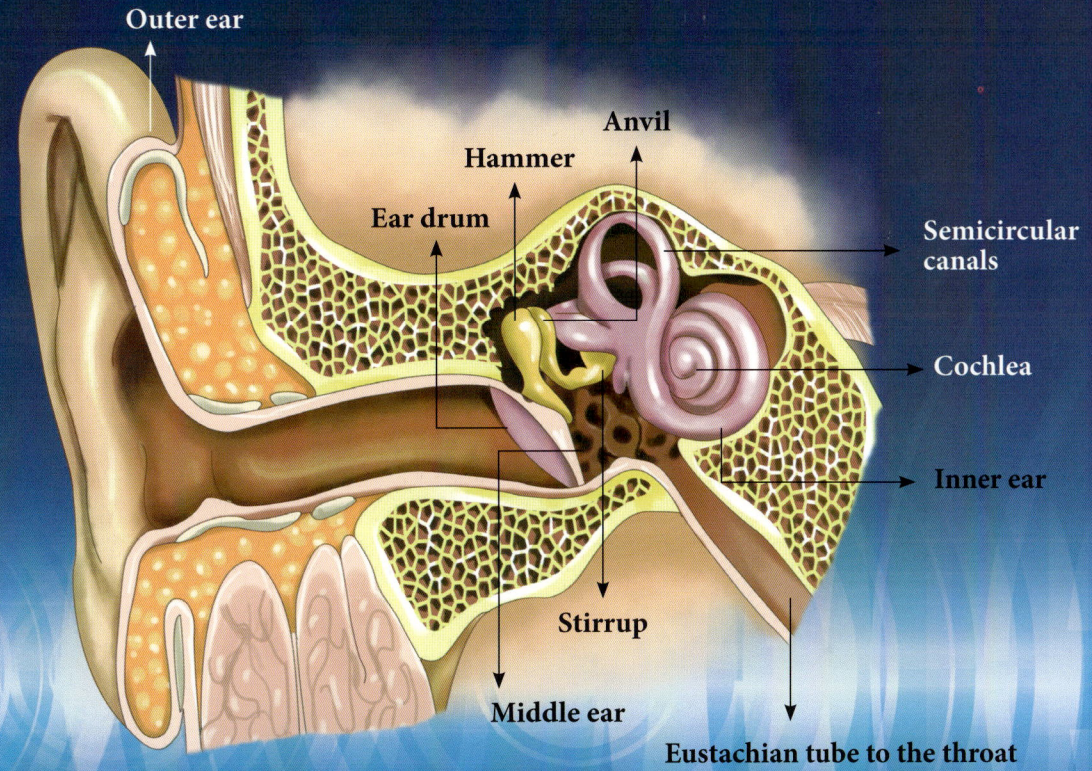

When animals talk
Herbivores, like deer, usually have big ears that capture more sound to keep it safe from predators. Want to hear like a herbivore? Cup your palms behind your ears and fan your fingers out. Notice how sounds suddenly appear louder? Horses, rabbits and kangaroos have ears that can twist around without moving their heads. This keeps the animals on alert against danger.

Like humans, animals use sound to communicate too. A rattlesnake shakes the end of its tail, creating a rattling sound. This sound warns other animals from coming closer. The low howling of wolves is carried on for miles and helps the pack stay together. Crows are known to have a complex language system. They make atleast twenty five different types of sounds. They growl, squawk, squeak, coo and rattle to communicate with other crows. They even make emergency calls when needed.

Energy watch Bats and dogs have larger hearing ranges than us, while frogs hear only between 100-2500 Hz.

ECHO, ULTRASOUND, AND SONAR

You've surely heard an echo. Standing in an empty room, when you said 'HELLO', you might have heard it repeated to you faintly. That's an echo of your voice. Echoes are similar to throwing a ball at a wall and having the ball bounce back to you. But, did you know echoes are more than just a fun past time?

Talking back: Echo
Like a mirror reflects light, sound gets bounced back when it falls on a hard solid object, like a wall. A softer surface, like cloth, will absorb most of the sound instead. This is why in a room full of furniture, curtains and carpets, there is no echo. But an empty room is a fun place to bounce your voice around.
You usually hear your echo a short while after you speak. The farther you stand from the reflecting surface, the longer the delay. If you shout at a hillside, your echo returns after few seconds. Sound takes time to travel forward and then the same amount of time to travel back to you. Hence, the delay. Infact, if this delay is less than 1/15 of a second, your brain will not distinguish the echo from the call. You need to stand atleast 17.2m away to hear your echo.

Ultrasound vs. Infrasound
Human ears can hear sounds only up to 20,000 Hz. Sound waves with frequencies above this are called ultrasound. A dog whistle produces an ultrasound call, which dogs hear but we don't. Ultrasound is most commonly used by doctors to look inside a human body. A machine transmits millions of ultrasonic sound waves into the body. The reflections and echoes of these sound waves generate a picture, which is projected on a screen. This technology is used to monitor the growth of babies inside their mother's womb.
Similarly, infrasound are those waves with frequencies less than 20 Hz. This is lower than a human ear's listening range. Infrasound is used to study earthquakes and search for underground oil reserves. Elephants are known to use infrasound for communication.

Sonar
Sonar stands for 'Sound Navigation and Ranging'. Sonar helps in locating underwater objects using echoes. Submarines and fishing boats send out a series of sound pulses, called pings. These pings vary in frequency from infrasound to ultrasound, and can be as powerful as a million shouts. Receivers pick up the returning echoes. The time taken by the echoes gives data to calculate the object's distance and shape.

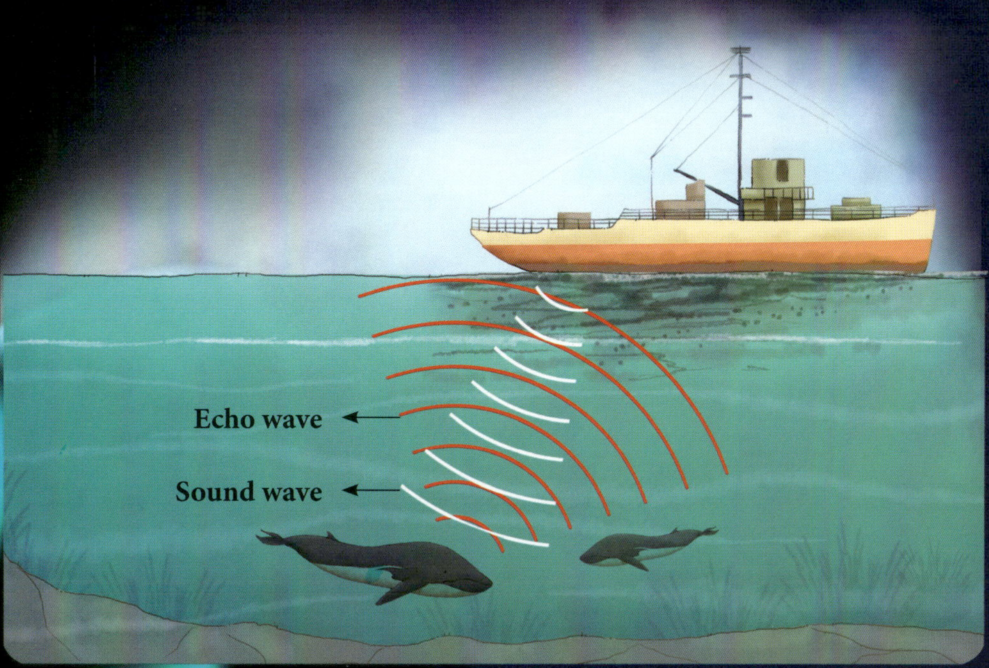

Echo wave ←

Sound wave ←

Sonar has been around in nature for longer than we began to use it. Dolphins navigate and search for food using echolocation. They create a series of clicking noises, called click trains, which are sounds at different pitches. The higher the pitch, the more details they receive.

Bats use 'echo' to see in the dark. They send out high pitched (ultrasound) calls and listen to the echoes to identify objects in their flight path. They can even spot something as small as a mosquito!

Sounds good

Sound travels at 343 meters/s. If an echo is produced 2 seconds after the call, how far is the reflecting object?

Answer: Approximately 343 meters away. In 2 seconds, sound travels a total of (343 x 2) meters to reach the object and return. Half this value is the one-way distance.

Energy watch — An echo is fainter than the original call since the reflecting surface absorbs some of the energy.

SOUND POLLUTION

Why should we care about sound pollution? Isn't sound an essential part of our lives? Sound helps us communicate and stay informed about our surroundings. It entertains us, soothes us, and even warns us against danger. However, sound can also become unpleasant and unwanted. And too much of an unpleasant sound is annoying, distracting and often painful to our ears.

What is noise?

Talking is necessary, but too many people talking together turns into noise. Noise is an unwanted sound; it disturbs us and prevents us from hearing wanted sounds. Loud songs blaring at night affect our sleep. Constant sound of construction interferes with normal activities, like talking on the phone. Excessive noise affects our hearing. Sound pollution is the highest form of pollution today, with cities generating the most of it. Mumbai is the noisiest city in the world, followed by Tokyo and New York.

Where does noise come from?

Noise broadly falls under two categories – manmade and natural. Animals, birds, wind, storms etc. are natural sources. Manmade sources include household gadgets, motorbikes, loud speakers etc.

Household sources: The churning of the kitchen mixer, the hum of the vacuum cleaner and the whir from the air conditioner are some of the noise generators at home. Televisions and music systems are common too. A pet dog, barking at every shadow, disturbs others in the area, especially at night.

Industrial noise: Construction sites, noise from heavy machinery and large exhaust fans in factories, noisy equipments like drill machines etc. Some industries mandate workers to wear noise-cancelling headphones to avoid health issues caused by noise.

Transportation: The inside of an aircraft may sound pretty quiet, but outside, it is deafening. Airport ground staff on the runway are exposed to 140dB in one go! City traffic adds greatly to noise pollution.

Am I adding to sound pollution? Tick the appropriate box for each question.

Do I?	Always	Sometimes	Never
Listen to the television so loud, my neighbours know what I watch.			
Play music at high volume so people outside my house /car can hear.			
Burst noisy crackers during festivals like Diwali.			
Talk loudly with friends at school and shout unnecessarily in the playground.			
Slam doors when I close them			

Which category has more ticks? Never, Sometimes or Always?
If you have more ticks in 'Always', can you think of ways to reduce the noise you generate?

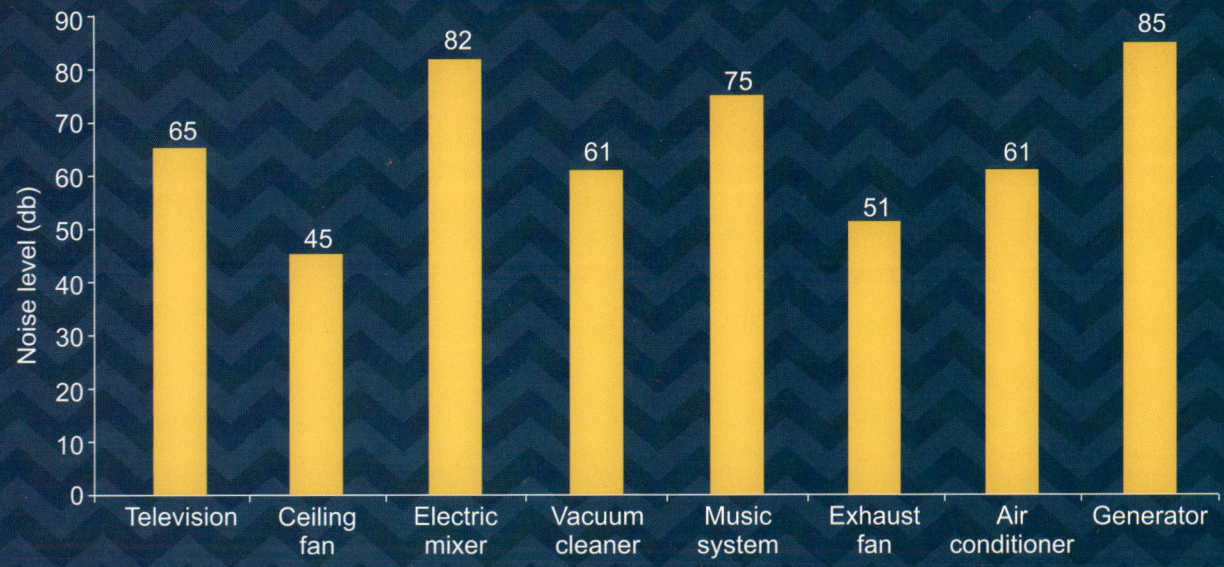

Noise pollution at home

Effects of sound pollution

- **Hearing:** Prolonged exposure to loud noises leads to a hearing loss. Any sound louder than 85 dB is harmful to ears. Traffic noise, an MP3 player at maximum volume, sirens, and firecrackers all fall under this category.
- **Animal life:** Noise from aircrafts and jets affect the normal ways of life for birds and bats. Loud tourist helicopters, flying over wildlife reserves, frighten elephants into stampedes. Excessive sonar in seas turn the waters noisy, and cause injuries and deaths to whales and dolphins.
- **General health:** Sound pollution increases anxiety levels and disrupts our daily life. Night time pollution leads to sleep disturbances, and is even worse than daytime pollution.

Make listening safe. Limit your time on your personal music system to one hour a day. Take frequent breaks from the time spent on noisy activities.

Sounds good

Quizzz?

Will putting up thick curtains protect your home against noise?

Answer: Yes, heavy curtains absorb sound to a great extent and reduce noise.

Energy watch — World Health Organisation recommends healthy noise levels at less than 50dB during the day and 40dB at night.

LET'S MAKE A **MUSIC BOX**

Music without musical instruments sounds incomplete, doesn't it? Guitar, flute, tabla, sitar are few of the popular instruments. Then there are some unique instruments, like the Theremin, which you can play by waving your hands in the air. Or the Glass Armonica, invented by Benjamin Franklin, and made of stacked glass bowls. With this simple shoebox guitar, let's make some music of our own.

To make a simple shoebox guitar, you will need
- A cardboard box with a lid
- Scissors
- A pointed needle to make holes
- Five long rubber bands
- Card paper
- Cardboard tube (optional)
- Wrapping paper, glitter glue, crayons, sketch pens (optional)

1. Start by cutting out a large circular hole on the lid. If you wish you can cover the box with a coloured wrapping paper.

2. Make five small holes in a straight line on either side of the sound hole, about 1" away. The holes should not go beyond the widest point of the circle.

3. Make similar five holes on two rectangular strips of card. These holes should be in line with the holes you made in step 2 on the lid.

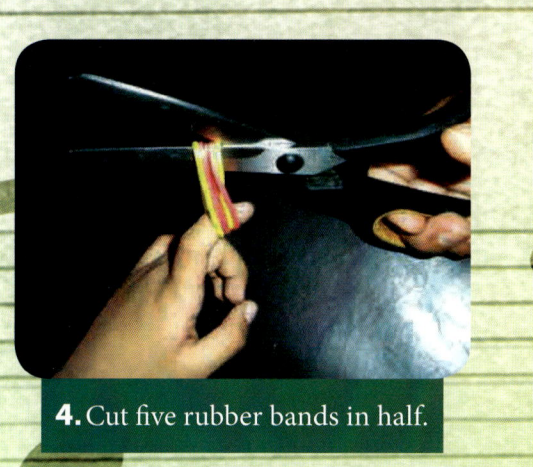

4. Cut five rubber bands in half.

5. String the rubber bands through the hole in one card and knot them. The rubber bands should not slip out of the holes when you pull them.

6. String the rubber bands through the holes on one side of the lid. Then stretch them and pass them through the holes on the other side. On the back, pass the rubber bands through the second card strip and knot them. Make sure the rubber bands are properly stretched out. You could vary the stretch on different strings to get different sounds.

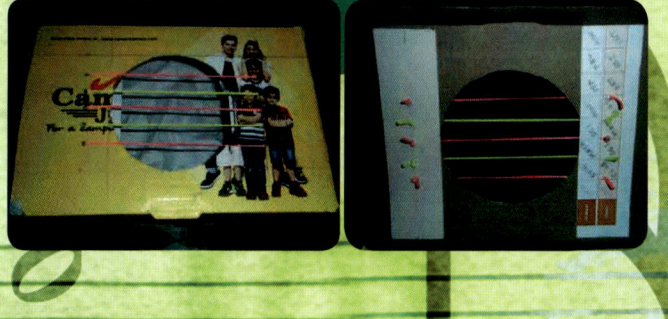

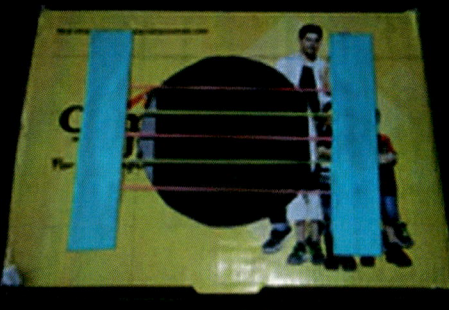

7. Stick two strips of card on the front of the lid, on top of the holes to hide them. The sound box is ready.

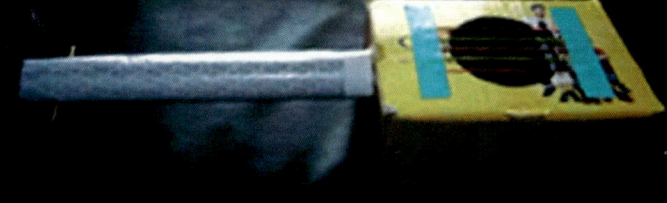

8. Stick a cardboard tube or roll to one end of the sound box to make it look like a real guitar. Decorate it using glitter glue, crayons or stickers. Strum along to your favourite song.

Energy watch

The sound box is what amplifies the sound from the guitar's strings and makes it musical and interesting.

ENERGY AND THE ENVIRONMENT

Every year we consume more energy than we did before. A lot of that energy gets wasted. Energy is not free, but importantly, using energy is not without an impact on our environment. While fossil fuels pollute the air we breathe in, wind turbines are noisy. Every type of fuel, whether renewable or non-renewable, affects us in some way.

Fossil fuels: Dirty sources of energy

Fossil fuels generate huge amounts of carbon dioxide, a greenhouse gas (GHG). This lead to global warming and deteriorate our health too. Drilling for oil disturbs land and marine animals. Oil spills contaminate the soil and water. Coal, natural gas, and oil have made our lives easier, but we pay a huge price in terms of health hazards and damages to our environment.

Renewables: Near zero pollution

Renewable energy are considered 'green energy', but they too affect our environment in certain ways. Large scale solar facilities require dedicated land space. Wind turbines threaten habitats of birds and bats. People living close to wind facilities complain of noise and vibrations. Dams, built to harness hydropower, flood the surrounding area. Flooding destroys wildlife, forest cover, and agricultural land.

Global warming Greenhouse gases envelop our planet and keep it warm. While a warm planet is good, excess of these gases is making our planet a little too hot.

Hope for a clean energy future

To reduce the impact of our energy use, we need to understand it first. Today, we can discover oil reserves with fewer exploratory wells, thanks to technology. Abandoned oil wells are plugged and the area is restored. However, non-renewable sources cause much more damage than renewable sources. Hence, it is important to move away from fossil fuels and adopt better, cleaner ways to meet energy demands.

    Bio fuels may emit GHG too, but the quantity they produce is far less than fossil fuels. Solar panels on rooftops and abandoned mining land reduce their negative effect on land and habitats. Although electricity is considered 'clean energy', its production may have affected the environment. Hence, energy efficiency is the need of the hour. Being energy efficient means using less energy to do tasks as well as before, or even better. Sunny skies, gusts of wind, gushing waters, plant residue, and heat from the earth promise clean energy future.

Become an Energy Star

We cannot avoid using energy. We need energy to cook, to light our homes, for transportation etc. However, our energy choices directly affect our health and our environment. Choose smart...

- choose energy-efficient appliances. Look for the Energy Star symbol when buying. Use LED light bulbs.
- choose to turn off lights, fans, air conditioners, televisions, computers, water heating, cars etc. when not required.
- choose to put food inside the refrigerator after it cools down.
- choose to recycle. Less energy is used to manufacture products from recycled materials.
- choose to walk, bike, use carpools, and public transports.
- choose a broom and mop, instead of vacuum cleaners.
- choose renewable energy, like drying your clothes in the sun instead of the electric dryer.
- choose cold water to wash clothes instead of hot water.
- choose to plant trees to shade your home.

Fact file

China consumes the most energy in the world, while Germany leads as an energy-efficient nation. India ranks third in consumption and stands at #11 in energy-efficiency.

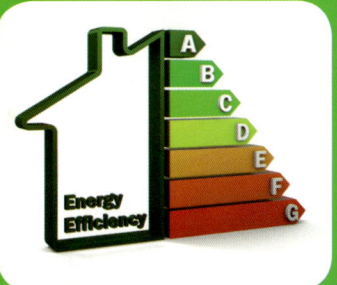

Which fossil fuel causes greater air pollution: diesel or coal?

Answer: Coal. It's the worst offender of all.

Energy watch One-fourth of the world's population lives without electricity. By saving energy you could help light up a home.

LET'S GET ACTIVE

1 Why is outer space dark and quiet?
....
....
....
....
....
....
....

2 VIBGYOR stands for:

V ___ ___ ___ ___ ___

I ___ ___ ___ ___ ___

B ___ ___ ___

G ___ ___ ___ ___ ___

Y ___ ___ ___ ___ ___ ___

O ___ ___ ___ ___ ___ ___

R ___ ___ ___

3 Specific heat capacity of Gold is 0.03 calories / g °C. Its melting point is 1064 °C. If room temperature is 30 °C, how many calories of heat is required to melt 1g of Gold at room temperature?

....
....
....
....

Choose the right answer.

What form of energy does food provide the human body?
a. Chemical energy
b. Mechanical energy
c. Electrical energy
d. Nuclear energy

What is one way in which physicians examine their patients?
a. By using a camera
b. By using a microscope
c. By using a video machine
d. By using an ultrasound machine

In which scenario below does mechanical energy convert into sound energy?
a. Rice cooker
b. Alarm clock
c. Flat iron
d. Oven toaster

Which statement is true?
a. Energy can be created
b. Energy can be destroyed
c. Energy is conserved
d. Energy increases over time

Which of the following is NOT an impact on the environment?
a. Fossil fuels emit large amounts of carbon dioxide gas
b. Wind Turbines generate noise pollution
c. Non-renewable energy sources have limited quantity
d. Coal mining requires cutting down trees to clear land area

Who am I?

I fall on the floor, I fall on walls
But, shine a light on me
and you won't see me at all... **9**

....

I am hot and I am light
At noontime, I am strong and bright
But you can't see me when it's night... **10**

....

I've been around hundreds of years
I come from OLD dinosaur bones.
You see me everywhere
But I could be soon gone... **11**

....

Sometimes I am gentle
Sometimes I am gusty
I can blow things away
and get you a little dusty... **12**

....

I am clean, I don't pollute
I generate lots of energy
But, Fukushima and Chernobyl show
Why I can be extremely risky... **13**

....

Name the following forms of energy.

14

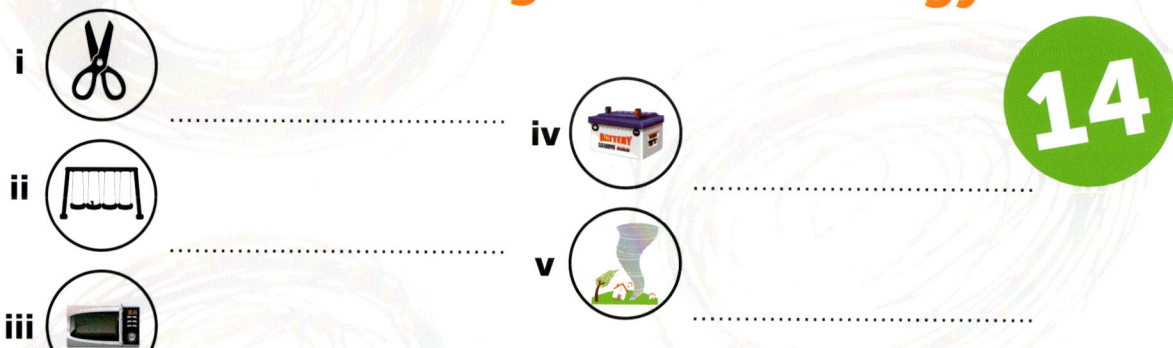

i  ....

ii  ....

iii  ....

iv  ....

v  ....

Fill in the blanks.

What are the units of measurement of:

i) Heat energy: J _ _ _ _ S and _ A _ _ _ _ _ S

ii) Temperature: C _ _ _ . _ _ _, _ A _ _ E _ _ E _ _
and K _ _ _ _ N

iii) Loudness of sound: _ E _ _ _ E _ S

iv) Frequency of a wave: _ E _ _ _

Answers

(3) 31.02 cal raises the temperature of 1 gm Gold by 1034 degrees

Choose the right answer: **(4)** a **(5)** d **(6)** b **(7)** c **(8)** c

Who am I?: (9) Shadow **(10)** Solar energy **(11)** Fossil fuel **(12)** Wind energy **(13)** Nuclear energy

Name the following forms of energy: (14) i) **Mechanical energy** ii) **Gravitational energy** iii) **Thermal energy** iv) **Chemical energy** v) **Motion energy**

Fill in the blanks: (15) i) Joules and Calories, ii) Celsius, Fahrenheit and Kelvin, iii) Decibels, iv) Hertz

GLOSSARY

Amplitude	Measure of the height of a wave. Stronger waves have higher amplitudes
Atmosphere	layer of gases that surround the earth.
Audible	loud enough to be heard
Chemical	substances which are used in factories, farms, and homes for a variety of purposes such as cleaning, painting, killing pests and maintaining vehicles.
Conservation	to preserve something from being destroyed
Echolocation	method of locating objects by determining the time for an echo to return and the direction from which it returns
Electron	subatomic particle with a negative charge, found in all atoms
Environment	the surrounding in which a person, an animal or a plant lives
Fluorescent	a substance capable of emitting radiation, especially visible light
Habitat	the place or type of site where a plant or animal naturally or normally lives and grows.
Molecule	the smallest unit of a substance that has all the properties of the substance. A molecule is made up of two or more atoms
Magnifying glass	a lens that produces an enlarged image of an object
Medium	an intervening substance, as air, through which something gets transported
Mirage	an effect caused by hot air, in which a sheet of water seems to appear in a desert or a stretch of hot road
Nucleus	the central part of an atom comprising of neutrons and protons
Opaque	not allowing light to pass through
Photon	a unit of electromagnetic radiation
Pollutant	any substance, like chemicals or waste products, that renders the air, soil, water, or other natural resources harmful or lowers their quality
Transparent	able to see through
Vacuum	completely empty space in which there is no air or other matter
Wavelength	distance between successive peaks or crests in a wave of sound, light etc.